ARCHAEOLINGUA

Edited by
ERZSÉBET JEREM and WOLFGANG MEID

Series Minor
14

JOHN CHAPMAN

TENSIONS AT FUNERALS

Micro-Tradition Analysis
in Later Hungarian Prehistory

BUDAPEST 2000

Printed with the support of the University of Durham,
the Centre for the Archaeology of Central and Eastern Europe

Cover illustration by Gina Stancu

ISBN 963-8046-29-5
HU-ISSN 1216-6847

© ARCHAEOLINGUA Foundation

All rights reserved. No part of this publication may be reproduced, stored in a retrieval system, or transmitted in any form or by any means, electronic, mechanical, digitised, photocopying, recording or otherwise without the prior permission of the publisher.

2000

ARCHAEOLINGUA ALAPÍTVÁNY
H-1250 Budapest, Úri u. 49

Word processing by the author
Desktop editing and lay-out by András Kardos
Illustrations by Yvonne Beadnell

Printed by Amulett '98 KFT Budapest

Contents

Preface 7

1. Introduction: research issues in Carpathian prehistory 9
   1.1 Three images from Hungarian prehistory. 9
   1.2 Research issues since 1982 12
      1.2.1 Traditional concerns and counter-cultures 13
      1.2.2 The processual agenda in Hungary 15
      1.2.3 Into the 1990s: advances in archaeological science 18
      1.2.4 Towards social archaeology in Hungary 21
   1.3 This book 24

2. Social approaches to the mortuary domain 27
   2.1 Introduction 27
   2.2 Social archaeology 28
   2.3 The mortuary domain 29
   2.4 Categorical analysis 30
   2.5 Dynamic nominalism 34
   2.6 Micro-tradition analysis in archaeology 37
   2.7 The sample 41
   2.8 Summary 42

3. The Late Neolithic intra-mural burials at Kisköre-Damm 45
   3.1 Introduction: Late Neolithic burial practices 45
   3.2 The site of Kisköre-Damm 46
   3.3 Previous mortuary studies and global rules 48
   3.4 Micro-tradition analysis 57
      3.4.1 Burial lines not related to houses 59
      3.4.2 Burial lines related to houses 61
      3.4.3 Personhood and group identities 66
      3.4.4 Tensions at funerals 69
   3.5 Summary 73

- 4. The earliest formal cemeteries in Hungary 75
  - 4.1 Introduction 75
  - 4.2 The Basatanya cemetery: background and previous analyses 76
  - 4.3 Global rules at Basatanya 81
  - 4.4 Micro-tradition analysis 83
    - 4.4.1 The Period I-II transition at Basatanya 84
    - 4.4.2 Period I grave lines 86
      - 4.4.2.1 Personhood and group identities 100
    - 4.4.3 Period II grave lines 101
      - 4.4.3.1 Personhood and group identities 117
    - 4.4.4 Discussion 119
  - 4.5 Summary 124

- 5. The Late Copper Age cemetery at Budakalász 125
  - 5.1 Introduction 125
  - 5.2 The Budakalász cemetery: background and previous analyses . . 126
  - 5.3 Micro-tradition analysis 130
  - 5.4 Discussion 156
  - 5.5 Summary 160

- 6. Conclusions 161
  - 6.1 Theory and method 161
  - 6.2 Results 163
  - 6.3 Wider implications and future research directions 166

- References 169

Preface

This book developed out of the discovery of a new kind of mortuary analysis, which I stumbled across in the course of writing about the fragmentation of human bodies and objects in another context. Soon after the discovery of the potential of graves organised in lines, I was asked to contribute a chapter to a book on agency in archaeology edited by Marcia-Anne Dobres and John Robb. The two topics began to mesh and, after much more research on agency and a re-analysis of the Hungarian data, the result is this short book.

I wish to acknowledge my debt to the University of Durham, who granted me a year's research leave during which some of the fundamental research for this work was begun. I am also grateful to Emma Blake, whose work on dynamic nominalism made a big impression on me and who kindly allowed me to use the material while still in press. Sam Lucy, Richard Hingley, Eszter Bánffy, László Bartosiewicz, Alice Choyke and Magdalena Seleanu all looked at parts of the text and offered valuable comments. Pál Raczky and Anna Endrődi gave me useful advice on the Budakalász cemetery and Jo Sofaer Derevenski has discussed the Tiszapolgár cemetery with me. I am also grateful to Simon Wyatt for making available his unpublished University of Edinburgh MA dissertation on the Tiszapolgár cemetery. I wish to thank Yvonne Beadnell for her excellent illustrations, Gina Stancu for the cover illustration and Erzsébet Jerem and her colleagues at Archaeolingua for much good advice during book production. I dedicate the book to Ludmila Koryakova, of late an inspiration to me.

1. Introduction: research issues in Carpathian prehistory

1.1 Three images from Hungarian prehistory

(1) It is dusk on the Great Hungarian plain, the largest lowland expanse in Central Europe. On the flattest parts of the plain, as in the Northern part near the river Tisza, anyone looking across the plain sees more sky than anything else. The only features disturbing the horizontal sweep of the skyscape are clumps of trees, the occasional group of one-storey huts and rare tells – mounds of settlement debris, which rise up to 5 m above the land surface. The people of this area have congregated to watch a mortuary rite – the burning of the houses on the tell of Late Neolithic Csőszhalom – after the death of a great community leader. For days now, kinsfolk and friends have carried pots, personal ornaments and their favourite flint tools to the houses and deposited them inside. On the last day before the firing, fuel is brought from around the tell and some of the few remaining trees are cut for firewood. The wood is piled up around the houses. Everything has been prepared.

The people stand around the foot of the tell as the ritual specialists form a procession and walk around the tell, chanting songs and throwing objects into the great circular ditches. After walking around the outside of the outer ditch, the procession crosses the entrance and moves slowly around the inside of the next ditch. There are five concentric ditches: with the crossing of each ditch, the procession moves closer to the central point of the mortuary ritual – the inner ring of houses, with their grave goods and their fuel. After the final chants have died away, the ritual specialist is handed a torch and she moves round the piles of brushwood, setting fire to each house in turn. Assistants fan the flames, which soon burn up and outwards, driving the worshippers back with their heat. The whole tell is ablaze and sparks fly out across the ditches to the amazed onlookers. Further away, other people – distant kin and non-kin – watch the immense blaze, as it colours the whole landscape and seems to reach into the sky. The burning tell can be seen from kilometres away and the stories begin of the great leader who has died and the honour which is done him by the size of the fire. The blaze continues through the night, tended by the ritual specialists and discussed by all the people of the area, who are connected visually to the tell by the destruction of its houses.

(2) It is the shortest day of the year – the winter solstice – not far South of the Csőszhalom tell. There is a half-metre of snow on the ground and the small

family who live in the farmstead near Bosnyák domb have been hungry for weeks. Two days ago, hunger fatally weakened the oldest member of the family – a 60-year-old woman who had been suffering badly from arthritis in her spine for some time. The burial is today and it is over a kilometre, across the deep snow, to the lineage cemetery by the creek, on the hill called Basa tanya. It has been difficult to move around the area to tell other kinsfolk the sad news, so the family does not expect many mourners to come to the burial. Towards noon, the group moves off, three of the younger members carrying the corpse. The children go ahead, trying to make a path through the snow but it is hard. After slow progress of an hour and a half, the family reaches the Basa tanya hill, to find only two old friends of their grandmother there. With so much snow, it is hard to find the farmstead's burial line but eventually a place is selected, near the South East corner of the hill, not far from a double burial of their cousin who died in childbirth two years ago.

The next problem is that the earth is frozen after more than a month of snow. It takes the adults enormous effort to break through the turf-line with their solitary Middle Copper Age copper axe: the antler mattocks were simply not strong enough. So a very shallow grave pit is dug to take the body. What else can be done ? – the risk of damage to the body is great but the ground is too hard. The oldest surviving family member chants a funeral dirge and the body is laid to rest. One friend places a small complete cup near to the head, while the family breaks three vessels and places fragments of each near the body. Finally, the children empty a bag of limestone beads over the body and the earth is thrown back into the shallow pit. After a few words of comfort and friendship, the groups move off towards their farms, trudging through the snow along the same route but without their dearest grandmother. The family members carry the fragments of the vessels, to remind them of their loss. When they reach their farm, the fragments are placed next to the fireplace, to honour the memory of the deceased.

(3) "There were always doubts about the couple who lived in the next farm, especially since their children died young and in strange circumstances. The man had moved from over the sand hill region to the Danube valley in search of good land and found a wife in our lineage. But when, one day, they were both found dead in the hut they had built just a few years before, it was evident that they had to be buried in the Late Copper Age community cemetery. So, the next day, a group from the nearest farms went to the farm and carried the bodies to the sand-hill overlooking the Danube, called Luppa-csárda. A good deep oval grave pit was dug in the sand, next to the double grave where their two children had been burnt and buried earlier that year. But the feeling amongst the

small group of mourners was that we had to signal the difference of the couple, so a big fire was made in the base of the grave, to purify the earth, just like we did with their children. It burned strongly for over an hour before the bodies were inserted. Our kinswoman was nearer the centre of the fire and was strongly burnt. The man's head was badly burnt. He was given no grave goods but the typical female costume – a necklace of Danube shells and shell beads – was placed upon her back. We filled in the grave but few people had come, despite the unusual fire. It was a quiet occasion, since there were no living relatives from their farm and not many had kept in touch from neighbouring farms".

Three images from Hungarian prehistory, each touching upon a period which we shall explore later in this book. These narratives are, of course, fictitious but based upon a particular reading of structured deposition of material culture which is arguably possible. Their historical accuracy is not a serious issue, since no narratives, scientific or otherwise, which we tell about prehistoric times can be said to be accurate. What these images do is to focus our attention on issues which we shall discuss in this book about the mortuary domain. These issues concern people as social actors – as carrying out a range of day-to-day tasks and occasional extraordinary practices which leave consistent material traces. As repeated activities leave more of a trace than single acts, it is likely that their material remains will be more visible in the sites and monuments which we investigate. I should emphasise that the conscious motivations and unconscious causes evoked in the narratives are not necessarily the only reasons behind such social actions. But there is an important link between social practices and structured deposition, which is found on many individual sites and monuments but is also present as a general structuring principle in the Hungarian Neolithic and Copper Age. The only way in which I wish to use such narratives is an exemplary use: they reveal something of the human background to structured deposition by presenting examples of social practices which result in the deliberate deposition of material remains which, at the same time, represent an act of categorisation, signifying an important aspect of the community's beliefs and practices.

This way of looking at the mortuary record, whether as the burning of houses on Late Neolithic tells or in Copper Age cemeteries, has antecedents in social theory, which we shall examine in more detail in the next chapter. But to make sense of the Neolithic and Copper Age periods which we shall be studying, it will be helpful to discuss an overview of recent investigations, in an attempt to characterise the main research issues of the period from the 1960s to the 1990s. The main part of this book is devoted to a presentation and analysis of the mortuary domain of the Hungarian Neolithic and Copper Age. A small number of sites will be considered in relatively great detail, while the vast majority of sites will perforce be totally ignored. It will thus be helpful in this chapter to set the few in the context of the many; in short, to attempt a definition of the main strands of the Hungarian Neolithic and Copper Age research agenda and see whether, or to what extent, the mortuary analyses which follow will in fact articulate with current research interests.

1.2 Research issues since 1982

The key topics of research in Neolithic and Copper Age research have seen a significant shift over the last 20 years. While the foremost traditional interests of prehistorians – new material, cultural determinations, relative chronology and typology – continue to dominate the vast majority of county journals, this is true to a markedly lesser extent in national journals and new books. Here, there has been a strong trend towards what may be summarised as "Hungarian processualism", while at the same time discussions concerning social archaeology have begun to become significant. But we cannot readily diagnose **new** interests, because interests in environmental, economic and social archaeology have a long history in Hungarian prehistory, reaching back to the research of János Banner (e.g., his study of the ethnology of the Kőrös culture:1937) and other key figures (Laszlovszky – Siklódi 1990). It is more accurate to refer to trends and shifts of opinion, as well as renewals of interest in fields of research seemingly long neglected.

In this chapter, it is intended to review the main developments in Hungarian Neolithic and Copper Age research over the last two decades. The survey begins with 1982, for three very good reasons: it was the date of the publication, in Hungarian, of János Makkay's synthesis of the Hungarian Neolithic and Copper Age (Makkay 1982); it was the year of publication, in English, of Andrew Sherratt's principal articles on the joint Anglo-Hungarian project in the Szeghalom region of County Békés (Sherratt 1982, 1982a-1983) and it was the starting date of the main Öcsöd excavations, directed by P. Raczky and destined to be a testing-bed of a new series of approaches to fieldwork and excavation (Raczky et al. 1985). These researchers defined or emphasised the major directions in Hungarian prehistoric research which have been followed largely to this day. The formulation of new research questions in the early 1980s led to the consolidation of "Hungarian processualism" – still a vibrant research tradition today.

1.2.1 Traditional concerns and counter-cultures

The general response amongst Hungarian prehistorians to the New Archaeology of the 1960s and its subsequent development into Processual Archaeology in the 1970s was one of suspicion. Laszlovszky and Siklódi (1990) have demonstrated how any kind of theoretical viewpoint, pro- or anti-materialist, was difficult to support in this period. While a pro-materialist position was difficult since it gave the impression of support for the socialist regime, an anti-materialist position was dangerous because of the attitude of the regime itself. Thus theoretical archaeology without theories (Laszlovszky and Siklódi's term: 1990) was practised by the majority of professionals through the 1960s and 1970s. The traditional concerns of the publication of startling new finds, their typological analysis, cultural attribution and accurate relative dating was the main means by which theoretical concerns were avoided. Just as sociology was not then practised as a University discipline, so the analysis of prehistoric society was regarded as a similarly dubious field by most practitioners. The most significant exception to this trend was the mortuary research of Dr. Ida Bognár-Kutzián, whose social reconstructions of the Early Copper Age community at Tiszapolgár-Basatanya (see Chapter 4) would scarcely have been possible in the absence of her own specific political circumstances.

A good example of the traditional concerns of Neolithic and Copper Age research is the classic summary of the Alföld Bandkeramik (AVK) by Nándor Kalicz and János Makkay, whose text was completed by 1963 but which was published in 1977 (Kalicz – Makkay 1977). Apart from brief sections on houses, pits, figurines, animal and fish bones and radiocarbon dates, the volume consists more or less entirely of the presentation of a site gazetteer and the primary ceramic data deriving from single-phase sites or site components. A major goal of the study was the seriation of the ceramic data into an Early-Middle Neolithic transition phase and two phases of the AVK – an early, widespread Alföld phase and a suite of later Alföld regional groups. The main explanation for diachronic changes in ceramics appears to be the passage of time itself, since there is no attempt to identify variables which were related to different pottery forms or decoration. The attribution of late AVK pottery to six regional groups, all of which were exchanging their own specific pottery to all of the other groups, represented a major typological achievement, which has, by and large, passed the test of time. The definition of a transitional Early-Middle Neolithic Szatmár I phase did not last so long, however, thanks to the realisation that it was a creation of two independent but stratigraphically mixed assemblages (Raczky 1988).

Such careful and immensely detailed cultural and chronological attribution dominated the study of the Tiszapolgár-Basatanya cemetery (Bognár-Kutzián 1963), leading to the definition of the main two Early and Middle Copper Age cultures, as well as several regional sub-groups of the ECA Tiszapolgár culture (further documented in Bognár-Kutzián 1972). Before the widespread availability of large numbers of radiocarbon dates, the 1960s and 1970s were the golden age of "cultural archaeology" in the Hungarian Neolithic and Copper Age, and led to the definition of many of the now-familiar cultural entities of the Neolithic and Copper Age.

The often short chronologies favoured in the Hungarian Neolithic and Copper Age meant that the prevalent mode of explanation of cultural change continued to be diffusion of innovations from more advanced areas – viz. "ex oriente lux" (Childe 1939). A good example of this viewpoint can be seen in the writings of János Makkay, whose social archaeology is written under the shadow of the Near East and Anatolia. Thus, not only was Makkay a staunch defender of the Jemdet Nasr origins of the infamous Tartăria tablets, excavated in a structured deposit on a Vinča tell in Transylvania (Makkay 1969, 1985) but he also attributed many of his other important structured deposits, from "Scherbenmachen" (the deliberate fragmentation of pottery: Makkay 1975) and blood sacrifices (Makkay 1983) to Oriental stimuli. Equally, the discovery of metal forks in Hungarian or Transylvanian Copper Age hoards led Makkay to a wide-ranging search for parallels all over the Orient (Makkay 1983a). The paradox of Makkay's work was that the more numerous and fascinating the discoveries he made of structured deposits in the Hungarian NCA, the less likely it was that they were inspired by some remote Oriental influence but rather the more likely that local Central European communities possessed a complexity which Makkay could not bring himself to accept.

In this period, two areas of Hungarian research led the way in Central and East European studies: archaeo-zoology and field survey. The archaeo-zoological research of the late Sándor Bökönyi, who for almost five decades published the most important faunal assemblages from the Hungarian Neolithic and Copper Age as well as from other countries and other periods (for an appreciation, see Bartosiewicz 1998; for a parallel approach to archaeo-botany, see Hartyányi et al. 1968; Hartyányi – Sz. Máthé 1979).

As Bartosiewicz (1998: 13) rightly observed: "Sándor Bökönyi's scientific activity filled a major gap between two historically separate geopolitical areas. The Carpathian Basin itself has served as an important link between the Near East and the rest of Europe since prehistoric times. He recognized the wealth of information within this rich cultural ecotone that had been explored only moderately by even archaeologists. Ever since Rütimeyer's first archaeozoological publication in 1861, this discipline has had its roots in German-speaking Central Europe. Relevant research in Hungary thus not only contributed important regional information to the field of archaeozoology, but also linked the traditional Central European school of archaeozoology and anglophonic prehistoric research in the Middle East" – and Yugoslavia.

Secondly, the early decision to make a systematic field survey of the whole of Hungary, under the project designation of "Magyar Régészeti Topográfia" (Hungarian Archaeological Topography), led to a series of county-based volumes, beginning with Ko. Veszprém in the 1960's (Torma 1969). In view of the increasingly destructive effects of deep ploughing in this period, the Topography was an important programme providing excellent data for settlement archaeology (cf. Choyke 1981). However, few Hungarian prehistorians availed themselves of the opportunity to develop their own approaches to settlement systems, these data bases often being utilised by Western Europeans.

Developments in both archaeo-zoology and field survey ran parallel to early attempts at another form of science-based archaeology – palaeo-environmental reconstructions. While Hungarian geographers, geomorphologists and geologists devoted much of the 1960s and 1970s to the understanding of the deep Pleistocene sequences in the Alföld Plain, less attention was paid to the Holocene environmental sequence. Thus, it was researchers such as John Nandris, and later his PhD student Krisztina Kosse, who began to integrate archaeological with modern and palaeo-environmental data (Nandris 1972; Kosse 1979; cf. also work on Southern Hungary: Chapman 1977). Thus, certain research trends were visible in the late 1970s which were strengthened and given emphatic local form in the 1980s.

1.2.2 The processual agenda in Hungary

The widespread diffusion of processualist theories, hypotheses and methods among European archaeologists had a cumulative effect in Hungary, whose Neolithic and Copper Age specialists became increasingly concerned with a wider range of interpretative models than was the norm in the previous decades. These new approaches to prehistoric societies constituted a response to new questions (e.g., faunal taphonomy, lithic raw material sources, etc.) as well as to questions formulated more precisely than in previous decades (the meaning of contrasting settlement structures, the impact of recovery techniques on excavated samples, the importance of long series of radiocarbon dates from tell sites, etc.). In the 1980s, rapid developments took place in local, Hungarian science-based archaeology. The prospects for science-based archaeology were strengthened by two appointments: Dr. Bökönyi was appointed to the directorship of the Budapest Institute of Archaeology, while leadership of its interdisciplinary section passed to Dr. Bognár-Kutzián, with the later assistance of Dr. László Bartosiewicz. These appointment coincided with an extension of the Topography, whose methodologies became more explicitly and carefully designed, leading to the possibilities of inter-survey comparisons (p.c., J. Laszlovszky).

There were thus two inter-linked main thrusts of research: (1) the science-based study of chronology, natural resources and the natural environment; and (2) the re-interpretation of prehistoric cultures through the synthesis of newly available science-based archaeological data. There was a certain tension between the research directions of culture vs. nature, which remains in the discipline in Hungary today.

As illustrative of the second research direction, one of the key works of the early 1980s was Makkay's synthesis of new trends in Neolithic research (Makkay 1982). The achievement of this book is the integration of traditional research concerns of chorology and chronology with the new science-based evidence for environment, settlement patterns, population estimates, exchange and social organisation. It is no coincidence that Makkay was, at the time, heavily involved in the County Békés / Szarvas volume of the Topography (published as Jankovich et al. (eds.) 1989). Here, Makkay developed ideas about settlement types, the relationship between tell locations and exchange routes and local two-level settlement hierarchies in the Late Neolithic (tells and farmsteads). On the basis of population estimates of certain large Late Neolithic tells, Makkay proposed the existence of local central places with key ritual functions, supported by coeval farmsteads producing tribute for their centres. The importance of ritual as an integrative function stands out as a key proposal, especially in the Late Neolithic. The notion of central places clearly had implications for the rise of central people, hence Makkay's proposals for the emergence of social hierarchies in the Late Neolithic and Copper Age. These developments were also related to local economic conditions, in which the domestication of plants and animals had provided the basis for highly productive and reliable subsistence strategies. Despite the importance of these local processes, Makkay continued to insist on the importance of external influences, both Oriental and Balkan, upon these cultural and social developments. But this diffusionist bias could not conceal the interest of a complex, multi-dimensional story of change, based upon a careful evaluation of the significance of cultural, ritual, economic, demographic and social factors. It was a loss to the wider European community of archaeologists that Makkay's (1982) volume was never translated into other languages.

In parallel with this cultural research, which utilised the conclusions of science-based archaeology but within a carefully-defined cultural framework, a second major contribution of 1982 was the integration of Topographic data into a broader study of socio-economic change on the Hungarian plain – the result of an Anglo-Hungarian project led by Istvan Torma and Andrew Sherratt. While Hungarian researchers also developed interpretative models based upon Topography data, as in Makkay's (1982a) study of Alföld Linearbandkeramik settlement patterns in County Békés, Sherratt' approach demonstrated a new interest in social archaeology and the vital importance of exchange at various spatial scales (Sherratt 1982, 1982a-1983). In these studies, Sherratt contrasted several spatial scales of analysis for the Neolithic, Copper Age and Early Bronze Age: regional settlement contrasts in Eastern Hungary by period (1982), detailed diachronic changes in settlement pattern in the Szeghalom area (1982a-1983) and intensive intra-site collection of a single site in each of the Early Neolithic, Later Neolithic and Early Copper Age periods (1982a). The results were integrated with palaeo-environmental data and combined with newly-acquired data on lithic raw material sources, supplied by Dr. Katalin Bíró, to produce an explanatory model of settlement and social evolution over four millennia of Hungarian prehistory (Sherratt 1983, cf. 1987 for more details on exchange). While later critiques of Sherratt's model have focussed on weaknesses common to many processual explanations (e.g., Shanks – Tilley 1987), the original publications provided an exemplary use of the social evolutionary approach.

Yet a third research focus was the excavation of the Late Neolithic low mound at Öcsöd-Kováshalom, which acted as a testing-ground for the application of many important methodologies and techniques at the time new to Hungary (Raczky et al. 1985). The pioneering combination of soil coring, intra-site gridded surface collection and trial trenching created the framework for the expansion of the project into a well-targeted full-scale excavation, using dry sieving and froth flotation to supplement careful hand excavation and the use of Harris matrices. The use of quantified techniques of analysis of the excavated material was innovative for Hungary and led to a sharpening of focus on the intra-site differences between the three major and two minor settlement foci, as with Dr. J.Rasson's calculation of settlement occupation time through the analysis of fine-grained pit stratigraphies.

Other significant changes in the 1980s included further evolution of the archaeo-zoological field pioneered by Bökönyi, through the application of more detailed sampling frameworks and the recognition of the importance of taphonomic factors of various kinds, as used by Bartosiewicz at Csabdi (Bartosiewicz 1984). A parallel development was the discovery through systematic fieldwork of more than 100 sources of lithic raw materials, led by Katalin Bíró and supported by Erzsébet Bácskay, Katalin Simán and Viola Dobosi. This pioneering research culminated in the immensely successful Sümeg conference of 1986 (Bíró 1986) and the establishment of a permanent reference collection of lithic raw materials, the so-called "Lithoteka", based in the Hungarian National Museum (Magyar Nemzeti Múzeum 1991). Another keystone in the creation of Hungarian science-based archaeology was the establishment of a radiocarbon dating laboratory in Debrecen, under the direction of the late Ede Hertelendi. This laboratory started to produce large numbers of dates for Hungarian Neolithic and Copper Age excavations from the mid-1980s.

An important staging-post on the road of Hungarian processualism was the Szolnok Late Neolithic conference, whose two major publications were the exhibition catalogue (Raczky 1987; German version: Meier-Arendt – Raczky 1990) and the conference proceedings, which appeared as volume II of the series *Varia Archaeologica Hungarica* (1989). For many non-Hungarian speakers, the exhibition catalogue was their first introduction to the rich material culture of the Late Neolithic tells. The catalogue is divided into two main parts: a synthesis of the Late Neolithic of Eastern Hungary, using the materials collected during a decade of "Hungarian processualism"; and summaries of the results of hitherto largely unpublished excavations on five Late Neolithic tells – Hódmezővásárhely-Gorzsa, Szegvár-Tűzköves, Öcsöd-Kováshalom, Vésztő-Mágor and Berettyóújfalu-Herpály. In the introductory chapter, Kalicz – Raczky (1987) discuss overall developments in settlement patterns and architecture, material culture, exchange, cult and economy, as well as chorology and chronology (with lists of new 14-C dates). The site chapters deal with many of the key issues first raised by Makkay (1982) – the relationship between settlement form and economic base, the existence of 2-level settlement hierarchies, the social implications of demographic estimates for tells and the trade and exchange of exotic prestige goods. There is also a strong emphasis on the evidence for cult within tell settlements, as well as the remarkable developments in houses, as at Gorzsa and Herpály, and shrines, as at Vésztő. In effect, these volumes summarised the principal empirical results of Neolithic enquiries in the 1970s and 1980s but they hardly took the conceptual basis of the subject further than the stage reached by Makkay in 1982.

1.2.3 Into the 1990s: advances in archaeological science

In the late 1980s and early 1990s, many of the earlier processual trends and approaches based upon scientific archaeology started to bear fruit. This was particularly true of the Hungarian National Museum's Lithoteka, with its increasingly valuable store of raw materials collected from sourced sites and petrologically analysed (Magyar Nemzeti Múzeum 1991). This data base allowed researchers such as Bíró (1988, 1992) to identify changing lithic raw material patterns of production and distribution in differing phases of the Palaeolithic, Neolithic and Copper Age in the Carpathian Basin. For Bíró, this research effort culminated in her Candidate's Doctoral thesis, recently translated into English, in which she makes a complex spatial analysis of the large lithic samples from Late Neolithic tells and horizontal settlements in both Eastern and Western Hungary (Bíró 1998). Social implications are drawn from the contrasts in the mechanisms of distribution between these two areas, with regional distribution centres in the Lengyel culture but not in the Alföld Plains communities.

Another development in Hungarian archaeological science – the 14-C dating laboratory in Debrecen – produced some spectacular results with the dating programmes for Neolithic tells and Copper Age cemeteries in Eastern Hungary. Long series of dates for Tápé-Lebő, Gorzsa and Szegvár-Tűzköves enabled the construction of a far more accurate chronological scale for the Middle and Late Neolithic than ever before (Horváth 1991; Horváth – Hertelendi 1994; Hertelendi – Horváth 1992). In the latest studies, this approach was extended to other tells, such as Csőszhalom, Öcsöd and Herpály, with exciting results (Hertelendi et al. 1998). New dates from the Zagreb lab for the Tiszapolgár-Basatanya cemetery also stimulated quite different interpretations about the cemetery (Benkő et al. 1987; Forenbaher 1993; see below, Chapter 4). Finally, the plotting of dates from all the different cultural units of the Eastern Hungarian Neolithic and Copper Age has led to a completely new picture of chronological development, in which the steady, unilinear replacement of one unit by another is itself replaced by a series of multiply overlapping ceramic distributions, with each major transition marked by a period of centuries in which widely differing ceramic fine wares are available for selection by local communities (Hertelendi et al. 1998; for conceptual basis of this new pattern, see Chapman 1981 for the First Temperate Neolithic – Dark Burnished Ware transition). Since the developments in absolute chronologies were markedly weighted towards Eastern Hungary, we are still awaiting similar results from Transdanubian Neolithic and Copper Age sites (but see Győr-Szabadrétdomb: Figler et al. 1997: 212). The application of inter-disciplinary approaches on a broader scale would be of particular importance for the continuation of the County Zala microregional research, where changes in the cultural, artifactual and settlement sequence have been integrated with the archaeobotanical and archaeozoological results (Bánffy 1995; Bökönyi 1993; Gyulai 1995; Bartosiewicz 1995).

In many ways, the greatest change in scientific archaeology in the late 1980s and early 1990s was the introduction into Hungarian prehistoric studies of GIS (Geographical Information Systems) and other forms of computer-based mapping and data retrieval programmes. GIS was pioneered at the same time as inter-disciplinary studies were making fast progress.The Hungarian microregional approach to settlement pattern research was an ideal place in which to test the integration of inter-disciplinary studies on an unified GIS framework. The first such project to be published was the first volume of the Gyomaendrőd microregional study (Bökönyi 1992) – which Bökönyi claimed (1992a: 8) to be the first large-scale research project in the history of Hungarian archaeology. The combination of studies in remote sensing, geophysics, palaeohydrology, geomorphology, pedology, physical anthropology, archaeozoology, malacology and archaeobotany in a single research volume made exciting reading. Benefits of scale were also to be seen in the first-ever total excavation of an Early Neolithic settlement – at Endrőd 119 (Makkay 1992; Bökönyi 1992b; Takács 1992). Perhaps the only area where Gyomaendrőd I did not fully succeed was the attempt to integrate data on a GIS. Two very different projects, coincidentally both based in North East Hungary, were to make a critical difference to the implementation of GIS.

One of these developments occurred in the mid-1990s in the context of a major Anglo-Hungarian inter-disciplinary research programme – the Upper Tisza Project (for summaries, see Chapman – Laszlovszky 1992, 1993, 1994, 1995; Chapman – Vicze 1996; Chapman et al. 1997). This programme was the first regional field survey project in Hungary designed with the aim of integration of all the research on a GIS. The project GIS co-ordinator, Mark Gillings, heroically digitised over 800 sq. km. of Hungarian 1:10,000-scale map coverage to provide a platform for the analysis of many different types of spatial and non-spatial data, including satellite imagery, on an ARC-INFO-based GIS (Gillings 1997, 1998).

Concurrently, another huge leap in the scale of Hungarian prehistoric research occurred in the context of the major motorway building projects of the mid- to late 1990s. The excavations of large open areas enabled prehistorians to investigate intra-site structure and differentiation on a scale hitherto unimaginable given normal funding constraints. The Eötvös Loránd University Department of Archaeology (whose name has now, significantly, been changed to the Institute of Archaeological Sciences) invested heavily in the personnel and the hardware to create a GIS laboratory for the storage and analysis of the data from the M-3 motorway programme in County Hajdú-Bihar, as well as other subsequent motorway projects (Raczky et al. 1997). Spectacular results were produced for the excavation of the Csőszhalom tell (Raczky et al. 1994; Raczky 1998) and, even more so, for the large horizontal settlement – Polgár 6 – adjacent to the tell (Raczky 1997, 1998). Since much of these data remains unpublished, it is premature to assess the implications for other aspects of theory and practice but the early publications suggest that major new avenues of intra-site investigation are beginning to open up.

These further developments of the scientific archaeology of the 1980s have increased the diversity and breadth of Hungarian prehistoric data, while allowing the integration of cartographic and artifactual data on GIS a reality (e.g., the new CD ROM of the database of M-3 finds from many of the main excavations: Magyar Nemzeti Múzeum 1999). The new-found focus on intra-site research has tipped the balance away from wide-ranging external relations to the causes of internal variability. Thus, a renewed interest in questions of social archaeology has come about in the last decade.

1.2.4 Towards social archaeology in Hungary

It is fair to say that, even in the 1990s, social structure was still not widely regarded as an important research issue. For many Hungarian prehistorians, the social and the ritual remained out of reach, at the pinnacle of Hawkes' (1954) ladder of inference, much more difficult to understand than the economic and the environmental. Paradoxically, so much of the data essential for a vigorous and fresh investigation of social structure had already been collected in the decade of "Hungarian processualism" – the material was at hand for a major change of research direction. We may heuristically divide the research field into burials, settlements and artifacts and begin with the mortuary domain, since it was here that Bognár-Kutzián (1963) produced her ground-breaking analysis of the Basatanya cemetery.

One of the rare research studies from Western Europe and North America dealing with Hungarian data was Susan Nacev Skomal's investigation of the Tiszapolgár and Bodrogkeresztúr cemeteries, which took social structure to be central to the problem of social organisation, although the study suffered from unwarranted assumptions about the nature of Indo-European influences (Skomal 1980). Skomal was concerned to identify social status in these cemeteries but proposed that Indo-European influences casued greater social differentiation than in local Neolithic communities. Somewhat later, I began an examination of Balkan Neolithic and Copper Age burial practices, in which a variety of approaches was used to elucidate social structure, including a study of the spatial attributes of burials, a contextual approach to artifacts used as grave goods and in the domestic sphere, and an analysis of the mutual categorisation of humans and artifacts (Chapman 1983).

An important Hungarian study was the research of Istvan Zalai-Gaál on Lengyel communities (Zalai-Gaál 1986, 1986a, 1988). Influenced by Skomal but not so convinced by the hints at social ranking, Zalai-Gaál studied the social structure of Kőrös, Transdanubian Bandkeramik and Lengyel communities through their mortuary remains in Western Hungary. Using a three-level nested spatial framework of graves, cemeteries and groups of cemeteries, Zalai-Gaál sought to define family grave groups using physical anthropology and the now less than reputable serological analysis. Focussing on the Lengyel culture, Zalai-Gaál made out a case for the use of Sahlins' (1972) domestic mode of production by nuclear families, within a matrilocal society best documented at Mórágy. Zalai-Gaál claimed that descent on the female side and social dominance of males were the two basic structuring principles in the Central European Neolithic from the Linearbandkeramik period onwards. It is curious how infrequently this truly remarkable study is quoted in the Hungarian Neolithic literature.

Much recent research in the mortuary domain has focussed on the Tiszapolgár-Basatanya cemetery (Sofaer Derevenski 1997, 2000; Chapman 1997: for full discussion of these articles, see below, Chapter 4). In other studies, I have developed a long-term analysis of burial practices in Eastern Hungary in the context of an analysis of social power (Chapman 1994, 1995). The striking contrasts in the form of the preferred dominant site in successive Neolithic and Copper Age phases led to the notion that contradictions in any given social formation could be re-addressed, if never completely resolved, through the selection of a different social arena in the next phase (e.g., Earlier Copper Age flat cemetery as compared with Late Neolithic tell).

This approach carried over into the settlement domain, where one of the main issues to emerge early on in Hungarian social archaeology concerned social elites and the origins of complexity. An early innovator here was Makkay (1982), whose view of tells as central places surrounded by satellite communities led to the search for central people in graves as much as in settlements. However, Makkay's site population estimates have recently been questioned using archaeozoological data (e.g., Raczky 1998). Since Makkay's very high site population estimates implied high levels of social complexity in the tell populations, this issue is not so clear-cut with the new, much lower estimates, nor is the evidence for a 2-level settlement hierarchy so convincing. Nonetheless, the regular discovery of Rondels, or enclosed sites, in North and Western Hungary has led to the re-emergence of the central place – central person hypothesis for the Late Neolithic, that person often regarded as holding specific ritual authority because of the structured deposition of unusual finds at most Rondels (Raczky et al. 1994; Kalicz 1998).

The early social arenas of power model (Chapman 1993, 1995) was integrated with a broader concept of landscape archaeology, in which key notions of timemarks and place-value were discussed in relation to different site and monument types in Eastern Hungary (Chapman 1997a, 1998). This research was based upon attempts to identify past communities' attitudes to time – whether cyclical or linear or a combination of the two – and the value of the places which they inhabited, with particular attention to the difference between tells and flat sites (Chapman 1989, 1998). The importance of cultural memory and the presence of the ancestors in tells are both implicated in the idea of tells as "ancestral homes". Parallel studies using computer visualisation and palaeohydrological modelling of Upper Tisza Project data led Gillings and Goodrick (1996) to postulate interesting reasons for Neolithic settlement locations, both for the majority of dry-land sites and for the floodplain exceptions. Most recently, Whittle's investigation of the emergence of the Neolithic in Southern Hungary has identified less reliance upon sedentary lifeways than in most other accounts – a conclusion which brings the earliest cultivators much closer to the indigenous foraging communities (Whittle 1998, 1998a).

The third area of research on social structure concerns the world of artifacts. I have already mentioned the research into what may be termed structured deposition by János Makkay, whose interpretations were based upon significant insights into the social organisation of Hungarian NCA communities, despite his diffusionist emphases (Makkay 1975, 1983). Such studies continued into the 1990s, as in Regenye's (1994) study of the Bakonyszűcs deposit, Raczky's (1994) careful analysis of the two Csóka (Čoka) hoards, set in the context of prestige goods hoarding in the Late Neolithic, and the study of two Csőszhalom ritual assemblages in burnt houses on the tell (Raczky et al. 1996). But the person who has devoted most research effort to the question of deposition in a great variety of forms is Eszter Bánffy – the Hungarian prehistorian most consistently concerned with a contextual approach to objects and whose approach is akin to postprocessual archaeologies.

Bánffy's most substantial contribution so far is the long article in which she analyses the depositional context of anthropomorphic figurines and a variety of other cult objects throughout Central, Eastern and South Eastern Europe (Bánffy 1990/91). Bánffy examines all the important contextual variables for the deposition of these objects – their excavation context, associations, function, the degree of structuredness of the deposition, the relationship between different artifact types, their form, completeness and degree of fragmentation. Bánffy identifies specific foundation deposits as constructional offerings, *pace* Makkay, but also recognises the paucity of shrines in her study region, with a far wider distribution of cult corners, offering pits and burial contexts. Bánffy has continued her contextual research in the context of her microregional studies in County Zala (Bánffy 1995), as well as in detailed studies of individual artifact types, such as the cult tables of the Lengyel culture (Bánffy 1997).

Among the more recent contributions to understanding the social life of artifacts is my research on fragmentation, in which deliberate breakage of quite high proportions of specific artifact types is the norm for much of the Neolithic and Copper Age of Central and Eastern Europe (Chapman 2000). While the full implications of the ubiquity of this social practice remain to be elucidated, similar research on Lengyel altar tables has been carried out by Bánffy (1997), while Kalicz (1998) has clearly taken fragmentation of figurines and fine wares seriously in his study of the Neolithic of Western Hungary.

It can be seen that the varied approaches to the study of social structure and complexity in Hungarian prehistory have progressed at uneven rates and rarely in complementary mode. This review of the main trends in the last 20 years of research in Hungarian Neolithic and Copper Age studies indicates that there is now a strong and healthy research field many of whose activities fall under the umbrella term "Hungarian processualism". We can expect to see much vigorous activity in this area in the coming decade, not least in intra-site studies and GIS-related spatial analysis. Other post-processual directions include the contextual approach and a range of Upper Tisza Project research interests which cluster around the question of personal identities – as characterised by the inter-relationships between persons, places and objects.

1.3 This book

How does this book relate to the research trajectories just discussed? This book is perhaps unusual for many Hungarian prehistorians in that it is concerned principally with **social** archaeology – the study of past forms of social organisation and structure as inferred from archaeological data. It is also unusual in that there is a comparative emphasis on the long-term sequence through the use of specific case studies restricted to the mortuary domain. The research tradition to which this work is perhaps most closely related is that of István Zalai-Gaál and his mortuary analyses of selected Lengyel cemeteries.

In this book, the case is presented for two alternative approaches to site-based and, by implication, comparative mortuary analysis which transcend the cultural historical and processual approaches most commonly applied to funerary data. These approaches are complementary and mutually supportive, permitting the archaeologist to theorise both the overall structures of the mortuary rules and the choices open to individual social actors. "Categorical analysis" is selected to give insights into the overall structure, while "dynamic nominalism" is utilised to examine agency in the mortuary domain. The use of this combination of analytical tools provides new insights into mortuary data from some of the best-known prehistoric cemeteries in Hungary. The aim of the diachronic presentation of the results from these analyses is to provide a deeper understanding of the later prehistoric sequence in the Carpathian Basin as a whole. Let us now retrace our steps and take a closer look at the main distinguishing characteristics of recent research in mortuary archaeology West of the Carpathians.

2. Social approaches to the mortuary domain

2.1 Introduction

The discovery of a well-preserved Late Neolithic body at an altitude of 3,210 m in the Italian-Austrian Alps brought home to a global audience the notion that individuals existed in the past (Spindler 1994). While the debates which raged about the cultural identity of Ötzi and the location of his home community hardly touched many people who were not archaeologists, it was the idea of Ötzi's personal identity that was so strikingly original. The general public seemed to need a face and body with real skin, instead of just the dry bones which archaeologists have excavated in their thousands, to make the past come alive in their imagination. The associated discovery of a wide array of objects with Ötzi stimulated people to imagine how Ötzi lived his life, or at any rate prepared for his last trip. After the identification of the sources of Ötzi's flints, and the spectral analysis of his copper axe, other puzzles still remain – why were his arrowshafts all broken?; what was the significance of the tattooing on his body?; and why was he carrying a large basket (or was it a drum?)?. The location of his body in a gully just above a huge cliff face also added to the drama of Ötzi's story – the place where he lay down (? for shelter and rest, or to die at the appointed time) tells us much about this early alpinist, surely with some practice in crossing and re-crossing dangerous Alpine passes from the lush valleys to the South.

People, objects and places – this is the holy trinity in which identity is developed. The relationship between each pair of terms has its own contribution to make to the creation and continuous re-creation of personal and collective identities in the past, as much as in the present. The relationship between people and places is reflexive, with the value of places, their sense of enduring identity, contributing strongly to what makes a person from that particular place (Chapman 1998). The relationship between people and objects is also reflexive, in that objects extend what it is to be a human and take upon themselves something of the characteristics of the person who reproduced them (Chapman 1996). The reason that certain objects are deposited in particular places is related to a third reflexive relationship, by which the fame of the place is increased by the kinds of objects which are brought there, used and ultimately deposited there, and *vice versa* (Chapman 1998). These three forms of relationship exemplify an approach to the past in which the renown of people, the reputation of objects and the fame of places are mutually constituted, as the principal means whereby identity is developed.

The theme of identity was one of the key archaeological research topics of the 1990s – and with good reason, considering the increasing uncertainties and complexities of a global system which pits immense structural forces against local communities. But archaeologists have for long struggled to separate out the multiple strands out of which identities are created. In settlement archaeology, it remains difficult to recognise the "fingerprints" of a specific individual form of settlement. Even in household archaeology, where Tringham recognises that a household must have contained at least one adult male, one adult female and children (Tringham 1991), it is difficult to attribute the tasks reconstructed from material remains to specific individuals, as well as to particular genders (for the crisis in task-based gender archaeology, see Spector 1991). Mortuary archaeology does, however, provide a possibility for dissecting the tangled web of variables contributing to the multiple and complex identities of the deceased. It is a major claim of this book that a detailed analysis of groups of burials within a settlement or a cemetery, rather than a consideration of the whole mortuary population as a single entity, provides the means of distinguishing traits representing different kinds of social identities. To understand why this is an important advance, we should consider, in overview, the main advances in social archaeology and how this has affected approaches to the mortuary domain. We shall then turn to two approaches which, it is hoped, will help to resolve some of the current difficulties with mortuary archaeology.

2.2 Social archaeology

The notion of social archaeology has been implicit in the very foundations of prehistoric archaeology as an academic discipline over a century ago. The archaeologists of the antiquarian period attempted to unravel material culture similarities and differences by the use of ethnic stereotypes. Kossinna's clearer formulations of "ethnic groups" based upon recurrent associations of bounded artifact distributions (Kossinna 1896, 1911) implied a tribal social structure whose identity was mirrored in the patterning of the finds. Childe took up and refined Kossinna's methodology, while soon rejecting his racial theories, and provided the lynch-pin of cultural archaeology for the first half of this century (Childe 1929). Although social theory was supposed to mark a prominent development in the "New Archaeology"of the 1960s and 1970s, many processualists continued to rely upon the social evolutionary views of Sahlins and Service. The upshot was a typological approach to social structure, in which prehistoric societies were bracketed in one of the four anthropological archetypes – "bands", "tribes", "chiefdoms" and "states" – and there were few satisfactory attempts to explain social change. Marxist analyses of the social contradictions which generated cultural and social change were limited to Western Europe (e.g., Spriggs 1984; Friedman – Rowlands 1977), often provoking strongly empiricist forms of archaeology in Central and Eastern Europe (Laszlovszky – Siklódi 1990). With the advent of post-processualism in the 1980s, the notion of "the social" penetrating deeply into all forms of practices led to a broadening of the interpretation of social structures, giving rise to multiple narratives about past social structure. Three forms of enquiry proved of particular relevance: the discovery of engendered archaeology (Gero – Conkey 1991), the formation of social identities through ethnicity (Jones 1997) and the ascription of social power (Mann 1986, 1993; Miller – Tilley 1985; Miller et al. 1991). In each approach, the significance of the individual was highlighted, often in contrast to the social group. Individual behaviour, or at any rate the behaviour of identifiable classes of social actors, took on a new, cumulative importance in relation to long-term social structures. The question arose of how individuals created their own cultural and social identities through dwelling in particular places and through the making of specific objects (Chapman 1997a). Questions of social agency in relation to structure were raised as a result of insights derived from Bourdieu (1977), Foucault (1984) and Giddens (1987, 1991). These insights were combined with an emphasis on social practices and the diverse ways in which the material residues of social practices formed the principal subject matter of the discipline (Barrett 1988).

2.3 The mortuary domain

The mortuary domain has long provided source materials for the analysis of prehistoric social organisation and structure. Processual analyses based upon the fundamental analyses of Binford (1971), Saxe (1970) and Tainter (1978) focussed on quantifiable variables such as energy investment, grave goods quantity and diversity and decision-making. Subtle ethno-archaeological analyses, such as that of O'Shea (1984, 1996) for the Early Bronze Age Maros group, differentiated horizontal (? lineage) variation from vertical social differentiation in a way rarely paralleled in other studies. In most of the analyses of the "rich" Varna cemetery, the "super-ordinate dimension" represented by Europe's first array of goldworking, was proof enough of a social hierarchy denoting chieftains at a regional if not an inter-regional level (Renfrew 1978, 1986; Gimbutas 1977; Lichardus 1988, 1991; Ivanov 1991). For the most part, the reflectionist approach remained dominant in processual analyses; the wealth and diversity of grave goods or some such surrogate measure, was a direct reflection of the social organisation and its constitutive social personae.

Post-processual analyses of the mortuary domain started from the belief that material culture was not necessarily a reflection of social structure but an active ingredient in the creation of social identities and the construction of ideological frameworks. The notion that the survivors were making statements about their society which may have differed from its actual form, rather than just representing it, raised many possibilities for mortuary analysis. The principal one was that individual differences amongst the living, whether of age, gender or social group, could be negotiated at the graveside as part of a continuing search for identity and dominance (Chapman 1996, 2000; Lucy 1998). These analyses underpinned the notion that the construction of social reality was a recursive process, in which those categories of people whose values became important were in turn supported by the objects and places which helped to define those same values. The process of valuation itself became an important item on the agenda of social archaeology (e.g., Bailey 1998). However, the tensions between structure and agency identified in the general field of social archaeology have been under-theorised in analyses of the mortuary domain (but cf. Parker Pearson 1999). It is with the aim of rectifying this omission that an outline is given of the following two approaches – categorical analysis and dynamic nominalism.

2.4 Categorical analysis

The method which I shall use to attempt to demonstrate this more complex pattern of people-object relations relies on the notion of artifacts as categories, as developed by Danny Miller (1985, 1987). For those readers wishing to move directly on to more empirical material, I shall summarise this discussion in the next paragraph before offering a more extended treatment.

A useful way of treating artifacts is to consider their cultural biographies, in which their life-histories are part of their cultural impact. This implies that each person who makes, owns or uses an object makes some contribution to the item's biographical story. In this sense, personal elements become part of the object world, as a way of externalising and objectifying people. It can then be seen that the value of an artifact is inter-dependent upon the value of the person most closely connected to that object. It further follows that a categorisation of objects can provide many insights into the way in which people divide up their cultural world, since objects play such an important part in the construction of that world.

Miller defines 'objectification' as the foundation for a theory of culture. For Miller, objectification is a process of development in which neither social nor cultural form is prior but in which both are mutually constitutive (1987: 18).

Basing himself on Hegel, Miller characterises objectification as a dual process whereby human subjects externalise themselves in a creative act of differentiation and in turn re-appropriate this externalisation through sublation (1987: 28). If objectification is the very essence of the development of the human subject, externalisation amounts to the creation of particular cultural forms, while sublation is necessary for externalisation not to be experienced by people as rupture or loss. This view of culture – where culture is the externalisation of society in history – can have no independent subjects, since they are reflexively constituted and is also, by the same token, an assertion of the non-reductionist nature of culture as process. Objectification implies that the process of culture must always include an element of self-alienation as a stage in its accomplishment, meaning that the process of culture is inherently contradictory! Miller emphasises that materiality plays a key role in the constitution of this contradiction (1987: 33).

Miller takes examples from social anthropologist Nancy Munn to provide concrete examples of objectification. In her earlier work on the Walbiri hunter-gatherers of Central Australia, Munn (1973) identifies externalisation as the projection of individuals in iconographic representations, coming to understand personal experience in relation to a set of landscape-based media which contrast people as social beings and certain overarching social relations. These projections are then internalised, which creates the individual's 'being' in relation to age, gender and social group. For the Walbiri, the particular medium of objectification is the landscape; the moral and social order is mapped onto a cultural landscape, which is 'naturalised' by being mapped in turn onto natural features of the geographical landscape. The properties of the mapped landscape in turn provide permanence, authority and massivity which can legitimate the social world. There is an interdependence between people and landscape in the creation of Walbiri culture.

In her later work on the Gawa islanders near Papua New Guinea, Munn (1986) examines the problem of constructing the self-image of a society in relation to outside groups. Here, objectification is a process of externalisation and sublation which depends upon inter-societal relations. The exchange relations between islands objectify the social relations between traders and others, based upon the way the Gawans 'invest' themselves in the act of creating an object for exchange. In this way, objects are creative of social relations, through a process whereby people self-alienate as an externalisation (the giving of a gift), only to have this aspect of themselves returned in a new form which accretes to itself the substance of the exchange (sublation as the receiving of a counter-gift). There is a similar interdependence between people and objects in the creation of Gawan culture.

In a similar way to Munn, I have previously examined the personalisation of artifacts in relation to the cultural creation of individual actors by enchainment and accumulation (Chapman 1996, 2000). Enchainment relies upon the direct relationship between person and object, in which inalienable objects were created out of persons and the exchange of these objects carried part of the previous owner with it to the next person, leading to a chain of social relations defined by material culture. In contrast, accumulation led to the strengthening of a new type of relationship between persons and objects which was in tension with traditional, enchained relationships. This was constituted by the loss of the direct relationship between person and object in favour of a representation of an abstract value, such as wealth, by the object now devoid of its most intimate personal connotations. In both ways of relating, artifact production, as one form of externalisation, and consumption, as one type of sublation, are deeply constitutive of emergent social orders.

Miller reinforces the importance of things when he states (1987: 105) that: "artifacts are simultaneously a form of natural materials whose nature we experience through practice AND the form through which we continually experience the very particular nature of our cultural order". Indeed, for Miller, material forms are part of the central order of cultural construction (1985: 205). But the material world, as an established environment for social action, always tends towards naturalisation, not because of reification but because of its very materiality; the way in which the material world acts to objectify a particular representation of society tends to favour that representation which reflects the interests of particular social groups as they pursue their social strategies (1985: 184, 192). Miller distinguishes between artifact 'categorisation' and 'classification'; while the latter refers to the secondary level of evidence, the very varied division of artifacts into groups by the producers themselves, the former represents the order imposed upon the world by the creation of a cultural order and may be used to study social and material relations. Thus categorisation enables the objects to integrate the individual within the normative order of the wider social group, where it serves as a medium of inter-subjective order inculcated as generative practice in some version of 'habitus' (1987: 129-130). However, while antecedent material culture may be normative, norms are but tools for fulfilling strategies and effecting change (Johnson 1989). Thus, the introduction of history into cultural creation allows for the possibility of cultural or social change.

This summary and discussion of Miller's ideas prompts the proposal of a linkage between objectification and agency. The forms of externalisation are produced against the background of a constraining yet enabling tradition (as structure), in which the naturalisation of the material world is drawn upon by powerful groups against competing traditions. But exactly because objectification is the essence of an individual's development, it is also constantly the source of individual action. Thus, for each individual, externalisation consists of actions taken by knowledgeable actors in a process of cultural creation. In this way, objectification as agency may be characterised as the main process by which individual self-identities are created through externalisation and sublation.

It is possible to identify two systems of categories – for people and for artifacts. Because of the limitations of the physical anthropological data, I am constrained for the most part to use only three age/sex categories: adult males, adult females and children. A current, widespread assumption is that the cultural construction of biological sex is used to "read off", in a relatively non-problematic way, the corresponding gender category. However, there is no unambiguous and independent manner of defining the results of the parallel process – namely, the cultural construction of gender ! Hence, much cemetery analysis dependent upon the anthropological sexing of human bones relies on a circular argument for the gendering of the persons whose bones are buried. Until this quandary can be solved, social constructivist hypotheses used in archaeological mortuary analyses remain problematic. Perhaps DNA analysis will be able to provide more reliable criteria for the sexing of human skeletal materials in the future. I cannot claim to have resolved this problem here – which leads me to define personal categories on the basis of biological age and sex rather than social age and gender.

If artifacts are simultaneously a form of natural material whose nature we experience through practice and the form through which we continually experience the particular nature of our cultural order (Miller 1987), it follows that a categorical analysis of the artifacts found with categories of people – age/sex categories in particular – should prove useful in elucidating the cultural order of the communities in question. We should recognise that artifacts presence the Other – they act as vehicles for bringing past time and past biographies (human and artifactual) into the present. Artifacts are, in short, insider stories – known to group members only – which carry within them the community's categorisation processes. Categorisation relies for its effect upon principles of **in**clusion and **ex**clusion. The categorisation of any society by age and sex represents the crucial differences by which that society is structured. So a key question for exploration is how material culture is used to make interventions about age and sex roles. The basic approach adopted here is to use the material culture deposited as grave goods as a way of assessing the degree of polarisation in age/sex categorisation – in itself, an indicator of age- and sex-based tension. The opposite to age/sex polarisation in categorisation is the strong overlap between categories of artifact deposited with different age/sex categories. It is possible that the objects shared between different age/sex categories may be more informative about the cultural construction of gender than artifacts exclusively associated with a single age/sex category.

But how does self-categorisation work? The social practice of categorisation is situated in a far wider set of human behaviour whose exact relationship to structure and agency has been hotly debated in recent years. I wish to examine the pragmatist approach known as 'dynamic nominalism' as an alternative to the more problematic structure – agency frameworks postulated by Giddens (1991).

2.5 Dynamic nominalism

The approach termed "dynamic nominalism" is based upon what philosophers refer to as "pragmatism". Again, I provide a brief summary of this approach in the next paragraph.

In recent accounts of the relations between structure and agency, such those of as Giddens or Barrett, a reflexive relationship between the two is posited, in which the existing social structure constrains individual action while individual action shapes and influences the long-term structure. These accounts can be criticised because they fail to break down the very opposition – structure vs. agency – which they seek to understand. Another way is to develop the idea of self-description through categorisation, in which individual or group identities are formed through a continual process of refinement and comparison with the past. Thus new forms of identity and new categories emerge through the very process of self-definition against an external Other, who is excluded from the identity in question.

The most general insight of pragmatism was summarised thus by Rorty (1989: 3): "About two hundred years ago, the idea that truth was made rather than found began to take hold of the imagination of Europe". The denial of "an external reality" is common to most pragmatists, who hold the social construction of reality to be essential to understanding the human condition (Goodman 1995). This view is also found in recent gender studies, both in anthropology (Connell 1987; Moore 1993) and in archaeology (Dobres 1995; Robb 1994; Sofaer Derevenski 1997). Such writers are quick to reject a homogenised, unified picture of culture or society which renders social relations invisible (Sofaer Derevenski 1997), arguing instead that culture is contested rather than shared and that social practice is an argument rather than a conversation (Ledermann 1990). In this view, the continuous construction of society is an object of strategy and, if it happens, is itself an achievement rather than a pre-supposed postulate of structuration theory (Connell 1987: 44).

This view privileges the gendered individuals and corporate groups who make and re-make their own history on a daily basis, struggling to create their own representations of themselves and find standpoints from which to promote their interests (Arsenault 1991).

However, the difficulty for social pragmatists is the reification of the dichotomy between structure and agency which they seek to abolish. Barrett attempts to collapse this dichotomy by linking 'agency', as the means of knowledgeable action', to structure recast as 'tradition' (1994: 5, 36). Here, "tradition becomes a necessary condition of agency, where such traditions are the structural conditions reproduced, monitored and re-evaluated in actions and speech" (1994: 36). In Barrett's archaeology of memory and practice, "traditions are the dispositions towards understanding which people routinely display by their actions" (1994: 95). People know the world they inhabit and they re-work that knowledge through active engagement with that world. However, action and structure (as tradition) in this account still stubbornly refuse to dissolve into a unity of process and a framework of knowledge. Are there alternative ways of deconstructing the recursive agency / structure opposition which would provide a means of relating the creative potential of things, places and people to each other and to explanations of prehistoric social change ?

The approach termed 'dynamic nominalism' is, broadly speaking, a form of agency theory, developed in the writings of Foucault (1973, 1979). The aim is to reconcile structure and agency within a single mechanism through the attribution of a more active role to identity. Hacking (1995: 247-8) defines the core notion: categories of people come into existence at the same time as kinds of people come into being to fit these categories in a two-way interaction. An example which Hacking draws from Foucault (1973) is the way that, owing to the development of new institutional forms of discipline and uniforms, soldiers in the Early Modern period 'became' different kinds of people from Medieval soldiers. If social change 'generates new kinds of people' (Hacking 1995: 248), this underlines the essential role of history in nominalism. This approach has recently been used in a study of Sardinian nuraghi by Emma Blake, who maintains that generative power of self-categorisation means that it is not only a type of agency but also a structuring device; it is a process which individuals engage in as well as a framework for other practices (Blake 1999). This means that agency and structure come together in the formation of identities, which may be described as the practice of self-description through categorisation. Identity, then, cannot simply be reduced to a function of habitus but is rather a way of coming to terms with the world and the Other. As Beaudry et al. (1991: 154) note, cultural identity is a public act of mediation between self and others, through any sign or object that allows a person to "make his self manifest".

At the level of the group, identities become a selection of defining characteristics, insofar as to define a group is to map its limits and define it in terms of what it is not. A key cultural resource to which selection is applied is the material world and the places where this is displayed; these storehouses of cultural resources (Barrett 1988) provide material for the re-writing of group origins, a process of locating the Other in the past (Blake 1999).

The self-definition of a group is a selection from one's own history and origins – a narrative of inclusions and exclusions.

This approach differs in two main ways from the agency theories of Giddens, Bourdieu or Barrett. First, in agency theory, agency and structure are distinct, while, in dynamic nominalism, self-categorisation can work only if structure and agency are coterminous. Here, structures are constituted by ingrained practices, which define self and group in quotidian action but are open to change. This position is consistent with Connell's (1987: 94) criticism of Giddens' ahistorical agency, namely that, where the link between structure and agency is a logical one, the form of the link cannot change through history. Secondly, whereas theorists such as Barrett see human subjects defining themselves through a continuous process of rediscovery of practical knowledge, Blake argues that self-definition channels the process of knowledge acquisition, providing actions with a description which is already part of the process of self-definition. Thus, people and groups are constituted by a reflexive historical process – the creation of categories of people which leads to the emergence of people who fit the new categories. For example, James et al. (1998) identify a structural category in society which they term "childhood" – an ever-present category in each society but whose membership is constantly changing.

It is also worth emphasising that the constitution of sex and age is not merely a matter of categorisation through comparisons with the past and the Other. Identity-formation is also at the same time a **process**, through which people grow in a historically contingent sequence of circumstances (for children, see James et al. 1998; for identity, see Jenkins 1999).

Hence, the self-categorisation of age and sex would have been constructed on the basis of cultural discourses, which may have emphasised gender similarity, ambiguity, multiplicity or binary opposition. Those individuals whose externalisation included the making of human representations contributed in a specific way to the construction of the normative order of the wider social group; the manner in which they gendered the representations is an indication of the importance of current gender similarities or differences. In this way, aged/sexed artifact categories integrated individuals into a normative social order, just as new cultural forms may have challenged or resisted those norms. Sofaer Derevenski (1997) maintains that, just as repeated group adherence to a repertoire of material forms and social practices lends gendered structure to society, so gender identity is the cumulative identification of elements from this repertoire. This is similar to Lesick's (1996) view that gender embodies the nature of experience of each person's 'material environment', which is constituted by tangible material forms; in this way, engendering becomes associated with a particular set of material forms. How can dynamic nominalism best be utilised as a source of insights in the material world of the mortuary domain?

2.6 Micro-tradition analysis in archaeology

Five socio-spatial groupings have been identified in the mortuary practices of the Neolithic and Copper Age (Chapman 1983). These are: (a) the burial of an individual under or in a house; (b) the burial of a group of people near a house – the so-called 'Household Cluster'; (c) the burial of kinship members in one or more small groups on unoccupied parts of a settlement; (d) the burial of corporate group members in small cemeteries (10s of graves); and (e) the burial of members of a large corporate group in large cemeteries (100s of graves). Most archaeologists have selected two basic units of analysis in the mortuary domain – either the single body or the complete cemetery (or all those graves in a cemetery belonging to a specific period or cultural phase) (Chapman 1983, 1994 and references). By far the majority of mortuary evidence from most periods of the Neolithic and Copper Age concerns the burial of complete bodies. From the perspective of personhood, the more complete the buried corpse, the more complete the statement about the deceased's social persona and the greater the potential it provides for the communication of social and cultural messages. The formation of personhood is related to group membership, expressed spatially in the burial location and socially as those with whom a person is buried. Equally, the larger and more-encompassing the corporate group, the more rule-bound the mortuary domain. Thus the expectation is that mortuary practices within larger entities offered more opportunities both to make statements about individual personalities and to differentiate individuals, especially along lines of age and gender, as well as defining group membership. It is a truism that the discovery of cemeteries – a rare site category in the Neolithic and Copper Age – is an indication of successful group formation and maintenance; dead lineages tell no tales!

Since processualist mortuary analyses, the archaeological study of larger burial entities – cemeteries or large groups of intramural burials – is predicated upon the selection of a decently large sample of graves and the analysis of the variability of the complete set of buried individuals. Analysis of this pooled sample often results in the identification of the global range of mortuary practices at a particular site and a set of 'rules' by which archaeologists may categorise the community in question. This type of analysis is based upon an amalgam of individual social acts – a palimpsest of all the burials that have survived in material form. This is indeed the 'standard' practice for archaeological analysis of mortuary data sets (this author included: Chapman 1983, 1996; for numerous other examples, see chapters in Beck 1995); it is the equivalent of the approach to multi-phase monuments criticised by Barrett (1994). In the case studies that follow, I shall use smaller groupings within the complete cemetery or intramural grouping and any existing directionality, as exhibited in lines of burials, to break down a 'global' approach to the mortuary palimpsest.

In this approach, I recognise the limitations of temporal control on the formation of the different burial groups, in contrast to the more secure spatial understandings. What we often know about the burial groups comprising a cemetery is the number of burials they contain, the age-sex categories of the deceased members together with associated objects, and the distance of the furthest burial from the centre of any nearby structure (e.g., a house). What we do not know about the burial groups is their chronology relative to each other and the time interval between the successive burials, whether within the group or between groups (days, weeks, months, years). We are usually also unaware of the proportion of the whole community given full burial (for an exception, see the excellent physical anthropological study of Mokrin: Rega 1996). The assumption is made that each burial was deliberately placed where it was in substantial knowledge, through oral tradition and/or personal witness, of previous burials in the same group and in general awareness of the evolving range of mortuary variability practised by the community as a whole. Mizoguchi (1993) has explored this issue in connection with the sequence of Late Neolithic / Early Bronze Age burials in Britain. He observes that, because all human social practices are situated in unique time/space contexts, people are never free from the consequences of what they did prior to their current action (1993: 223). In becoming 'routinised', this repeated action constrains social freedom for new action. What separates past action from present decisions is social memory traces, which become an authoritative resource capable of manipulation by leaders who, in some sense, control the community's past (1993: 233). The mapping of the newly-dead onto the places inhabited by the ancestors is thus a deliberate social strategy for expressing a kinship calculus, a socio-spatial categorisation of persons with their complex cultural identities (for discussion of the location of barrows, see Barrett 1990). It thus becomes possible to take each burial group within the total site mortuary sample as a micro-tradition to investigate (1) the ways in which the micro-tradition is supported or challenged with each successive interment and (2) the relationship between the 'global' or community tradition and the 'local', burial group micro-tradition. The assumption is made that, in a linear grouping of burials, the burials are indeed made in the sequence that can be inferred from the final order and that no burials are inserted into the order at the 'wrong' time. It is also assumed that placing the newly-dead in a particular group indicates close kinship links.

One of the most secure ways of controlling for the sequence of burials is the selection of burials which are unambiguously laid out in lines. There are many such cemeteries in prehistoric and early historic Europe, not least in the Early Medieval period of North West Europe (Lucy 1998). In Germany, the name of this type of burial is "Reihengräber" (literally, 'row cemeteries') (Schülke 1999). Yet, as far as I can establish, no-one as attempted a mortuary analysis in which the grave line becomes the principal unit of analysis (p. c., A. Schülke; S. Lucy). The optimal analysis of this kind, which I propose to call "micro-tradition analysis", requires the establishment of five preconditions:

1. the presence of the minimal number of graves necessary to constitute a line
2. the identification of a line of burials from a cemetery plan
3. the unidirectional deposition of burials in the line from one end of the line to the other
4. the identification of which end of the line was the first burial and which the last
5. the placing of the vast majority of burials in a cemetery in lines rather than in another manner (e.g., non-linear groupings, pairs of graves or isolated graves).

Clearly, some of these preconditions are easier to satisfy than others. Given the size of most prehistoric cemeteries, it is arguable that the minimum number of graves held to constitute a line is three. The attribution of graves on a site plan to lines rather than to another spatial layout is fundamental to the analysis; if there is not a reasonably good agreement on this between several specialists, the analysis is liely to produce ambiguous or incorrect results.

It may be hard to demonstrate convincingly that the graves in a recognisable grave line were laid down in a single, unbroken order. However, several kinds of information can mitigate this problem. Three useful classes of spatial information include: (a) the distances between each pair of graves; (b) the presence / absence of inter-cutting graves; and (c) the consistency of grave orientation. Regular spacing of graves suggests that the time elapsed between burials is not so great and there is some above-ground, perhaps organic, marker of existing graves. Uneven inter-grave spacing makes it less likely for unilinear continuity of burial, especially with a big gap between graves. It may be that a gap of more than twice the mean length of a grave means the line is broken. The greater the consistency of grave orientation, the more likely it is for adherence to a grave line, other spatial factors being equal. The social point is that there are often good reasons for the establishment of a specific direction for a grave line. Once the direction of a grave line is established, there is social value in the maintenance of that line, which helps to establish the identity of the next burial relative to the existing grave line as well as to confirm the identities of previous burials in relation to the newly-dead. It may never be possible to prove that, in a line of equally spaced graves, grave C was laid down out of time/space order (i.e. after A but before B) but leaving a clear gap for an eventual burial B. But the complexities of future spatial planning for a death that has not actually yet happened seem inherently far less probable than the unidirectional burial of the newly-dead in an established grave line.

The question of the start of a line of burials is by far the hardest methodological question to solve. Barrett (1990) has shown how the sequence of barrows can be established within certain parameters, often using the fine chronology possible with Bronze Age metalwork as grave goods. With single Neolithic or Copper Age burials in grave lines, utilisation of grave goods is not so productive and the best opportunity is usually the identification of a relationship with an earlier structure or burial. In such cases, the existence of earlier burials or other structures can be convincingly related to the putative start of an early burial line. Selection of sites where such a relationship to ancestral features can be shown to exist would be an important factor in research design.

In summary, there are several pre-conditions which favour a more successful operation of a micro-tradition analysis. Careful selection of sites will be essential in an attempt to maximise the potential of the prehistoric mortuary domain for such a type of analysis. It is also clear that many of the classes of information used for interpretation of the existence or continuity of grave lines will be specific to the local context.

The establishment of an appropriate site with well-defined grave lines of defined directionality opens the way to several kinds of analysis. First and most obviously, the analysis of each grave line provides information about the local micro-tradition and the manner of initial definition and subsequent modification. Secondly, comparison of different grave lines allows the identification of what may be termed group traits – those elements of burial mode or grave goods which are either common in a grave line or unique to a specific micro-tradition or both. Group traits help to create a form of corporate group identity (extended family?, lineage?, other?) within parts of the total burial population and may be usefully compared with the global rules for the total population. Thirdly, comparison of the burial of individuals of varying age/sex within a single micro-tradition can indicate which burial mode or material culture is drawn upon to emphasise differences in age-based or sex-based identities. In addition, where individuals of similar age-sex character are found in the same grave line, differences in mortuary treatment may be used to establish the ways in which individual identities were underlined. Fourthly, one may compare the ways of creating such identities in each micro-tradition with all other micro-traditions – an analysis which is also valuable in contrasting local agency with the global rules "valid" for the total burial population. Hence, micro-tradition analysis has a high potential for the establishment of contextually-based differences in at least four kinds of identity – individual, age-based, sex-based, corporate group and whole-cemetery.

2.7 The sample

In this book, I wish to examine the micro-traditions of three prehistoric sites from Hungary. I begin with the Late Neolithic open settlement of Kisköre-Damm on the middle Tisza (Korek 1989) (location: *Map 1*). The intra-mural burial sample from this site was selected for four reasons: (1) the excellent publication provides good-quality archaeological and physical anthropological data; (2) the moderate sample size makes this a useful test case for a first application of the method; (3) there is a well-defined earlier burial line which crosses the settlement area with a concentration of intramural burials; and (4) most of the burial lines stand in some form of relationship to what are assumed to be contemporary Tisza houses.

I then turn to the largest cemetery from the earliest period in which cemetery burial was common in Hungary – the Earlier Copper Age cemetery of Tiszapolgár-Basatanya, in the upper Tisza area (Bognár-Kutzián 1963) (location: *Map 1*). I have already discussed the importance of this cemetery in the annals of Hungarian prehistory (see above, pp. 13-14); there are obvious benefits for a re-analysis of material which has been intensively studied in the past. But there are three additional positive factors about the Basatanya burial sample: (1) there is some evidence for small-scale burial prior to the layout of any Copper Age grave lines; (2) there is internal, spatio-typological evidence which supports the seriation of grave directions; and (3) many of the grave lines are clear and unambiguous.

Finally, I examine a published sample of about 25 % of the Late Copper Age cemetery of Budakalász – Luppa Csárda, near the Danube just North of Budapest (Soproni 1956) (location: *Map 1*). In some ways, this is the most

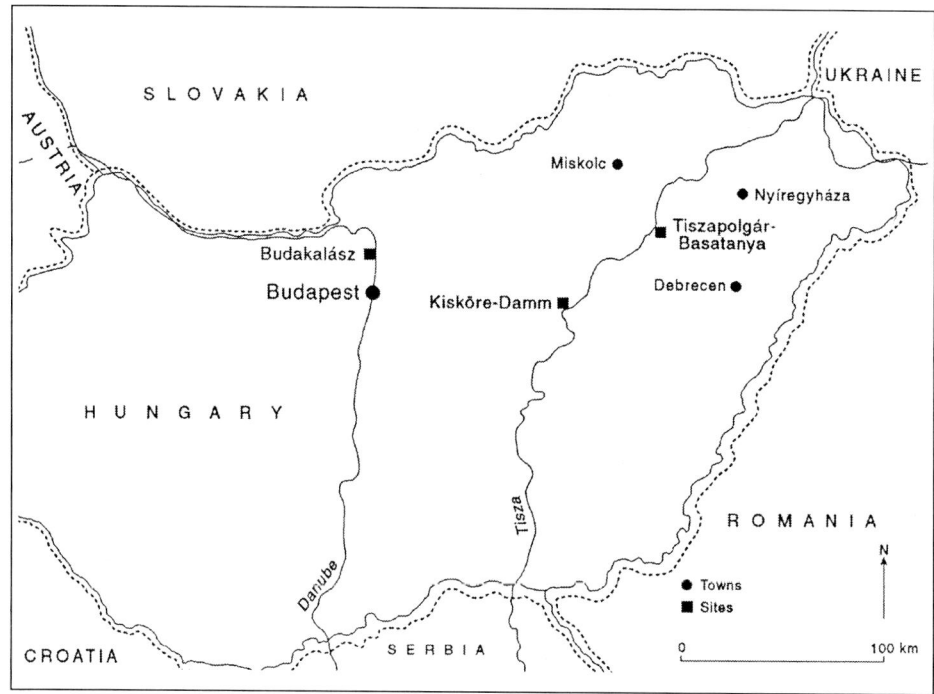

Map. 1. Location map of sample sites.

problematic sample of the three, since the published sample is incomplete and the physical anthropological classification is vestigial. In addition, there is no evidence for prior structures or burials in the part of the cemetery under study and therefore the directionality of the grave lines cannot be defined. Nonetheless, there is strong spatial patterning in the layout of the graves and the archaeological data are well recorded. In sum, the inclusion of Budakalász broadens the interpretative possibilities of the study, while missing the strong methodological advantages of Kisköre and Basatanya.

2.8 Summary

To summarise, the perceptual schemes used in cultural representations are a vital part of the community's categorisations of the material world and their place in it. The actualisation of the potential of the material world for making statements is as important a part of agency as any other cultural act, providing the framework for self-description through objectification and sublation, as well as the ongoing creation of personal and cultural identities. A dynamic nominalist approach is used as a framework for the cultural creation of identities. Because the selection of the medium of objectification is central to social power struggles, this can lead to a great variety of material strategies through which self-identity is constructed and proclaimed. Representation, then, is a key part of externalisation and hence central to human agency.

The mortuary domain is used here to explore the implications of this approach through the means of a new approach – micro-tradition analysis. The concept is based upon the selection of grave lines rather than individual graves or the whole cemetery as the principal unit of analysis. Each successive burial in a burial line is part of a micro-tradition (the burial line) which grows and has the potential to change with each new burial. The burial preceding the latest interment structures the selection of burial practices bestowed on the newly-dead but there is also a choice of alternative burial modes and grave goods for the latest burial. In this way, the burial ground is a meeting point for the structures which have been formed by other burials in each grave line and the social action which represents the choice of actual rite for the latest burial. Hence, micro-tradition analysis provides a way of integrating a dynamic nominalist approach with a contextual study of the mortuary domain.

3. The Late Neolithic intra-mural burials at Kisköre-Damm

3.1 Introduction: Late Neolithic burial practices

The first flowering of the tell tradition in Hungary dates to the Late Neolithic, c. 4500-3900 CAL BC. Even then, fewer than 20 tells and tell-like sites are known from eastern Hungary, their number being exceeded by usually larger flat sites (Kalicz – Raczky 1987). The height of the tells – 3-4 m – indicates intensive building and re-building, with rubble from earlier houses flattened and re-incorporated into new houses, a material strategy for the incorporation of ancestors into the world of the living (Chapman 1997a). The houses are well-built, comfortable, full of life, fertility, furniture and fittings and possessions, including pottery, figurines and other ritual paraphernalia. There is little doubt that the domestic arena of social power is salient on Hungarian tells. Because of their small numbers in the Alföld, each Late Neolithic tell assumed a greater place-value, in a wider social setting, than in tell-dominated Bulgaria (Chapman 1991).

The principal single burial mode in the Late Neolithic is intramural inhumation, found on tells and within flat settlements. All the recently excavated Late Neolithic tells boast numerous burials of partial or complete inhumations, usually of articulated skeletons, on unoccupied parts of the tell but, without exception, outside the houses (Kalicz – Raczky 1987; Chapman 1994). Tell burials indicate ancestral continuity in the realm of the living, as is demonstrated by the frequent direct superpositioning of houses directly on top of earlier structures (Bailey 1996). The first signs of gender differentiation occur with tell burial, with females lying on their left side and males on their right, both with predominant body orientation of SE-NW (Raczky 1987a), just like the preferred Middle Neolithic burial orientation and the orientation of many tell houses. However, this depositional polarisation is rarely reinforced by gendered grave-good differentiation. It is also in this period that the earliest hermaphrodite figurines occur (e.g., the famous throned figure from Szegvár-Tűzköves replete with breasts and penis: Korek 1987: Fig. 70), in the same occupation horizon as the gender-neutral "sickle-god" (Csalog 1959).

3.2 The site of Kisköre-Damm

The best-published example of intramural burial on a Late Neolithic settlement is Kisköre-Damm (Korek 1989: 23-45, 74-124). Area excavation exposed part of a Middle Neolithic (AVK: cca. 5200-4500 CAL BC)) settlement, a flat, unenclosed Tisza settlement with intramural burials and an Early Copper Age occupation (cca. 4500-4100 CAL BC). Four, possibly five, AVK burials were identified through the contracted inhumation rite which typifies all of the known Middle Neolithic burials on the Hungarian Plain (Kalicz – Makkay 1977; Chapman 1983, 1994). Only one of these five burials contained grave goods – Grave 12, with a grey rounded bowl characteristic of the AVK (Korek 1989: 41). By contrast, 31 Tisza burials were documented on the basis of the extended inhumation rite which is typical for that period on the Plain; in addition, many of the graves contained Late Neolithic Tisza pottery. Despite the small sample size, Korek's excellent recording of the mortuary features permits a wide-ranging analysis of the structure of the intramural burials. Skeletons were aged and sexed by I. Lengyel on the basis of both morphological and serological data. The two identification methods produced a high degree of consistency, enabling a differentiation into children (aged 1-16), young adults (aged 17-35) and mature adults (aged 36 and over). Other age-sex categories were not discussed in the physical anthropological report. The rich cultural data sets include the depth and orientation of graves, the cranial orientation, the costumes and grave goods found with the burials and the elemental materials used in costumes and grave goods.

There are two cultural antecedents with spatial meanings relevant to the Kisköre burials. The first is the line of earlier burials, made in the earlier AVK period (cca. 5200-4500 CAL BC), which bisects the area of the Tisza community excavated in 1964-66 (*Fig. 1*). An important issue is whether these four graves (nos. 10-13), with a single outlier (no. 22), represented 'distant ancestors' for the earliest Tisza occupants or whether the location of the graves remained unknown in the Tisza period (cca. 4500-3900 CAL BC). Two spatial and one cultural points are germane here. Spatially, the greatest concentration of Late Neolithic burials lies close to the AVK graves, yet no later burial disturbs the AVK graves. Culturally, the only Tisza burials with the typical AVK orientation – quite different from the typical Tisza direction – lie close to the AVK burial line. This amounts to the probability that knowledge of the earlier burials was transmitted in oral tradition, much in the manner that Küchler (1994) has outlined the mental mapping of previous land use in the decision-making of New Ireland groups for the establishment of new house-sites. The focus of Tisza burials close to the earlier ancestors betokened a series of strategic acts of reverence towards the long-dead and the cultivation of roots within the ancestral soil.

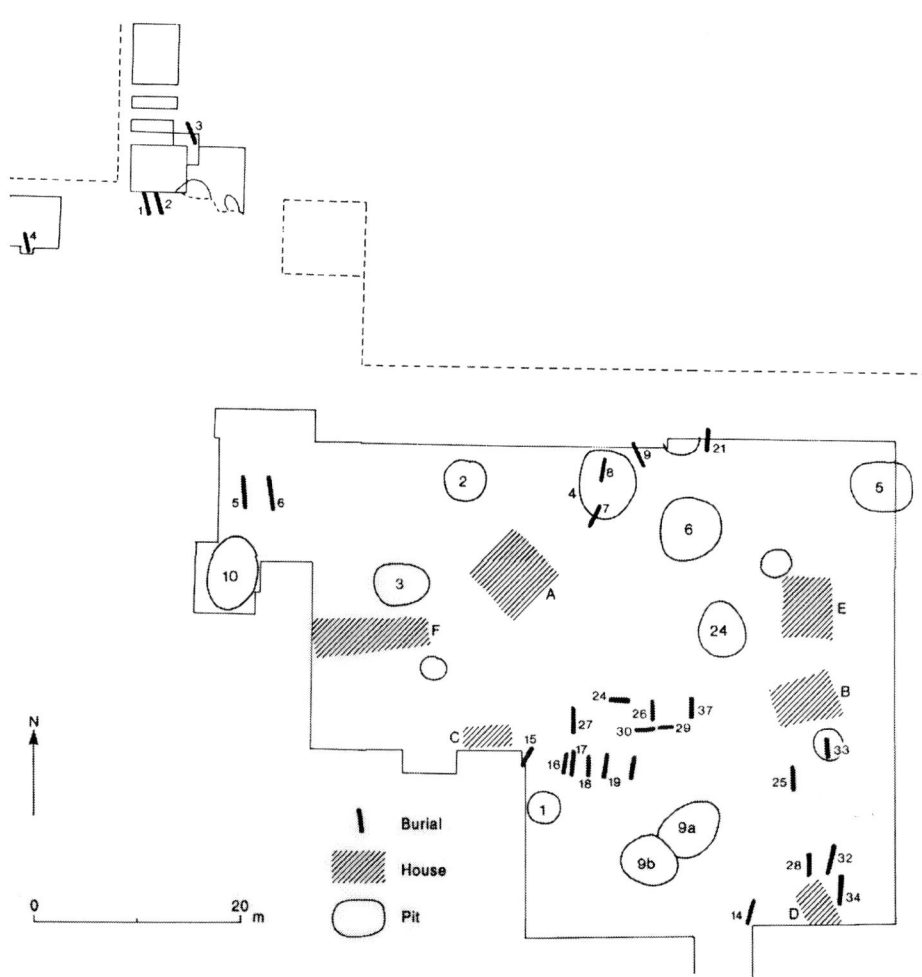

Fig. 1. Plan of selected Middle and Late Neolithic features, Kisköre-Damm.

The second antecedent feature at Kisköre was the six Tisza houses which can be identified at Kisköre. The assumption is made here that each burial group found near a house (*Fig. 1*) postdated the building of that house, as part of a Household Cluster. Since Late Neolithic houses tended to be more spacious and permanent than earlier Neolithic houses, it is assumed that the Kisköre houses formed a major structuring element in the settlement, a counterpoint to the 'permanence' of the buried ancestors, a group of features in which the materialisation of the domestic ideology was powerfully expressed and which in turn provided a range of material culture to be drawn on in the mortuary domain. A further three burial clusters were not associated with a house, a negative relationship which remains problematic. Are these groups any less 'domesticated' than the groups near houses? Equally, there is an unexplained absence of burials close to two large houses, E and F.

3.3 Previous mortuary studies and global rules

Before we examine each individual burial group, it is important to define the 'global' range of community mortuary practices. Since the excavator did not make a detailed analysis of the mortuary data, I shall present an abbreviated version before summarising the conclusions in eight points (for original analyses, see Chapman 2000, 2000a).

As in the case of Late Neolithic on-tell burials, a clear rule for grave orientation was established and, for the most part, closely adhered to in all grave groups. The predominant orientation was SE-NW, with the head to the SE. This rule was followed in all the male burials, 10/11 female burials and a majority of children's burials (7/11). One female was buried on a W-E orientation, while the rule was least strictly applied for children's burial, as if to emphasise their different status. Unlike the on-tell pattern, skull orientation was not only varied but also showed no correlation with gender. All age-sex categories used each of the three orientations – left, right and upwards. The combination of grave and skull orientation produces great variability: eight categories in the 22 graves which present complete information. As suggested above, male graves showed the least variation, children's graves the most.

The depth of graves was reported for 28 of the 31 graves: eight males and ten females and children respectively. There was no obvious patterning in the age-sex categories, with clear overlaps in the standard deviation of each mean.

However, when the burials are divided into those with and without formal grave pits (2/10 for children, 4/10 for females and 5/8 for males), those male burials in formal graves are significantly deeper than those without grave pits (0.90 m cf. 0.65 m). There are also significant differences in the depths of burials with and without costume (males – 0.98 m cf. 0.64 m; females – 1.05 m cf. 0.72 m; *contra* children 0.75 m cf. 0.93 m). Were it not for the reversal of the statistic for children's graves, this variable depth may have been thought to have some taphonomic significance for differential preservation of costume in deeper graves. There is no clear patterning in the comparison of burials with and without grave pits and bodies with or without costumes.

The objects deliberately introduced into the graves may initially be divided into three classes of finds – complete items, fragmentary items and sets of items. All categories of grave objects are represented in the first class, while sherds and fragments of shell bracelets comprise the second class; the third class usually comprises elements of necklaces or bracelets. All three classes combine to produce another sub-division into two broad categories: elements of costume adorning the body and separate grave goods unattached to the body or to clothing. At this juncture, it is worth recalling Sørensen's (1997) distinction between cloth, clothing and costume – three strategic levels at which items of dress may be studied. Here, the excavator of Kisköre has identified many items of clothing and decoration from the position of objects on the skeleton – in particular, the disposition of sets of beads (e.g. Korek 1989: Taf. 23). Korek has also identified the presence of shrouds from soil discolouration and cloaks of animal pelts from the presence of animal claws (1989: 23-45).

The first analysis focuses on elements of clothing and grave goods. No fewer than 37 categories of object have been deposited in the Kisköre graves – 20 elements of costume and 17 categories of grave goods. It is interesting to find that relatively few costume categories are shared by all three general age-gender categories (Table 1): each relates to a different part of the body, whether the all-over body covering or shroud, the red ochre crusted onto the skull, the necklace made of limestone beads and perforated red deer teeth, or the *Spondylus* shell bracelet.

Table 1. Grave good categories by age/sex categories, Kisköre-Damm.

Costume		Adult Female	Adult Male	Children
Head-dress	limestone beads	-	*	-
	head scarf	-	-	*
	limestone & deer teeth	*	-	-
Head	crusted red ochre	*	*	*

Necklace	limestone	-	-	*
	limestone & deer teeth	*	*	*
	limestone & fired clay	*	-	-
	limestone, clay & deer teeth	*	-	-
	limestone, *Spondylus* & deer teeth	*	-	-
Shirt	limestone beads attached	*	-	-
Cloak	animal pelt with claws	*	-	-
Arms	limestone bead bracelet	-	*	-
	shell bead bracelet	-	-	-
	complete *Spondylus* bracelet	*	*	*
	fragmentary *Spondylus* bracelet	*	-	-
Hands	bead rings	-	-	-
Belt	limestone & fired clay	*	-	-
	beads	-	*	-
Shroud		*	*	*
Placed grave goods				
Shell	perforated *Spondylus* shells	-	-	*
	twisted shell beads	-	*	-
	Dentalium / freshwater beads	*	*	-
Bone	cattle vertebra	*	-	-
	pig femur	*	-	-
	polisher	-	-	*
Flint	blade	-	*	-
Red Ochre	lumps	*	*	-
Limestone	beads	-	-	*
Pottery	whole rounded bowl	*	*	-

	whole dish	-	-	*
	whole beaker	*	-	-
	whole flowerpot	*	-	-
	fragments of rounded bowl	*	-	-
	fragments of amphora	*	-	-
	unidentif. fragments	-	-	*

The differentiation of adults from children is not strongly stressed, with less than a quarter of object categories found exclusively with either adults (3 cases) or children (6 cases). Unlike Copper Age cemeteries such as Basatanya (see below, pp. 80-81), there are no object categories which are shared between both children and adult females or between both children and adult males. In fact, the main difference emphasised by grave goods and costumes alike is the difference between adult females and adult males, with 14 categories exclusive to women and nine exclusive to men. Women tend to be buried with a wider variety of necklaces, composed of a greater variety of materials, while most male costume consists of sets of either limestone or shell beads. No metal artifacts are found at Kisköre and the only flint blade is found with an adult male burial. By contrast, the great majority of pottery is found with women – a trait paralleled in the exclusive female use of fired clay beads in jewellery sets. Furthermore, the rare deposition of animal bones is restricted to women's burials.

Finally, the only fragmentary objects are found in women's graves – whether sherds or bracelets. It is thus readily apparent that, at the categorical level, femaleness is betokened by a far wider range of grave goods and elements of costume than is either maleness or childhood. Nonetheless, because the elements of costume are predominantly sets of items, such as beads, the individual items may be found with males, females and children, even when the combinations are distributed more exclusively.

The relatively secure identification of age- as well as sex categories for both females and males at Kisköre allow the comparison of social identities as created through object association at different stages of the life-cycle (cf. Sofaer Derevenski 1997, 2000). While the difficulties of age attribution should not be minimised (Cox 1996), the ranges derived from the combination of morphological and serological analyses provides the basis for a simple tripartite categorisation for females – children (up to 15 years), younger adults (16-35 years) and mature adults (36 and over). Since no male children have been identified, male categorisation is restricted to younger and mature adults. However, bodies such as the female in grave 3 have been aged at 23-40 years (cf. the male in grave 14, aged at 36-52 years), so no clear-cut age-grades can be assigned in this analysis. The results are, however, encouraging and indicate two trends in the treatment of deceased females of varying ages. While females of any age-group can be buried without costumes, younger adults tend to have more complex costumes than either children or mature women, both in terms of complexity of costume elements and variety of element combination (Table 2).

Table 2. Burial costume by age/sex categories, Kisköre-Damm.

Grave	A/S Category	Costume
4	child	head dress + necklace + bracelet
37	child	none
27	child	none
24	child	necklace
26	child	none
30	child	none
29	child	red ochre on skull
28	child	necklace
33	child	none
1	younger female	necklace + 2 *Spondylus* bracelets
3	younger female	none
7	younger female	belt
21	younger female	head dress + necklace + shirt + bracelet
31	younger female	none
17	younger female	red ochre on skull
5	mature female	shroud
6	mature female	cloak + necklace
16	mature female	none

20	mature female	none
34	mature female	necklaces
35	younger male	bracelets
15	younger male	none
25	younger male	none
36	mature male	shroud + necklace + bracelets + ring
8	mature male	none
9	mature male	necklace + bracelets + belt
32	mature male	ochre on skull + head dress
14	mature male	none

The pattern for grave goods is that mature women are far more likely to have larger numbers of more varied grave goods than either children or younger adults, since the majority of the two last-named have no grave goods at all. The trend for males is even stronger: mature male bodies have more varied costumes and are accompanied by a wider range of more varied grave goods than younger males. Hence, while both mature females and mature males tend to increase their range of associated grave goods over younger persons, younger females and mature males betoken more varied costumes than other age cohorts.

The costumes and grave goods at Kiskőre are made of six classes of material – animal (e.g., the bone artifacts or perforated deer teeth), plant (the shrouds), earth (the pottery), mineral (the lumps of red ochre), stone (the flint or limestone beads) and shell (the beads and bracelets). It is interesting to note that five of these materials are represented in the four object categories common to all age-sex categories (red ochre on skulls; limestone beads and deer teeth on necklaces; shell bracelets; and plant-derived (? flaxen) shrouds). It seems probable that access to, control over, and use of, such basic elements of the natural and cultural worlds of the Kiskőre community was not limited to any particular age-sex category, even though there is some variety in the classes of materials from which object categories exclusive to age-sex classes were made. The only spatial variation in the occurrence of the materials is found in Line 5, where the children's graves are furnished with objects of stone and mineral only.

The next analysis concentrates on the age-sex distributions of the whole-body image of the Kiskőre skeletons – the mortuary costume. Nine elements have been identified by Korek in his meticulous excavations: ochre crusting on the head, the head-dress, the shroud, the cloak, the necklace, the shirt, the bracelet, the ring and the belt. The emphasis on visibility of display is reflected in the upper body distribution of most of these costume elements, with the sole exception of the full-length shroud and cloak. Sixteen burials betoken evidence of costume, ranging in complexity from one element to combinations of four elements. Thirteen different costume combinations are found in only sixteen burials; only two costumes are replicated in different age-gender categories (both adult women and children). This is a sign of the importance of the fluidity of social processes of identity-formation in the Late Neolithic, as much as the significance of complex, cross-cutting material culture categories.

The same principles of variability are demonstrated by the distribution of materials used in the creation of the Kisköre mortuary costumes (Table 3).

Table 3. Burial costume by raw materials by age/sex categories, Kisköre-Damm.

Raw materials	Adult Females	Adult Males	Children
Stone	-	-	*
Shell	-	*	-
Plant	*	-	-
Mineral	*	-	*
Stone/animal	*	-	*
Stone/earth	*	-	-
Stone/mineral	-	*	-
Stone/earth/shell	*	-	-
Stone/earth/animal	*	-	-
Stone/animal/shell	-	**	*
Mineral/plant/shell	-	-	*
Stone/animal/shell/plant	*	-	-

Twelve combinations of the six basic materials are found in sixteen burials; only one combination of materials is found twice in the same age-sex category and only three material combinations are ever repeated. This result is clearly influenced by the small sample size but, nevertheless, is still testimony to the notion that social identities are still relatively fluid and are created through permutations of similar elements rather than through fixed, binary oppositions.

The next analysis attempts to integrate the Kisköre mortuary costumes with the grave goods deposited. In view of the uniformly good preservation of the Neolithic graves, it would be hard to argue that there is not a genuine difference between those graves where bodies are costumed (n = 17) and those where they are not (n = 14). The question arises whether costumed bodies have systematically different sets of grave goods from bodies without costumes. It is evident that this is the case (Table 4): most grave good categories are far more likely to be found with a costumed body than an un-costumed one, with two exceptions.

Table 4. Grave goods in costumed and uncostumed graves, Kisköre-Damm.

Grave good	Costumed bodies			Uncostumed bodies		
	Female	Male	Child	Female	Male	Child
Pot	***	*	*	*	-	-
Bead	-	*	*	-	-	-
Shell	*	**	-	-	-	-
Flint	-	*	-	-	-	-
Polisher	-	-	-	-	-	*
Red ochre	*	-	*	-	**	-
Animal bone	**	-	-	-	-	-
No grave goods	***	*	***	***	***	****

The only example of a bone polisher was found in an un-costumed child's grave, while lumps of red ochre are equally likely to be found with either class of body. This finding suggests that the costuming of bodies and the deposition of grave goods is equally closely related to the creation of social identities and that each process is self-reinforcing, allowing the formation of more complex, cross-cutting social personae.

The analyses of the Kiskőre mortuary domain indicates several cross-cutting trends in the formation of 'personhood'. While most deceased are buried along a specified orientation, the combination of skull and grave orientation produces great variety in all age-gender categories. This is also true for the elements and raw materials comprising mortuary costume. The older the age of death, the more diverse the grave goods accompanying both females and males and the more elaborate the male costume (cf. female costume, which is most elaborate for younger adults). Variations in grave goods and costumes emphasise categorical differences between males and females far more than the differences between adults and children but these differences are expressed not through binary opposition but through the selection of different combinations of elements common to all age-gender classes. It is in this sense that the development of personhood at Kiskőre can be described as relational rather than oppositional.

The conclusions of these analyses may be summarised in eight 'global' rules:

1. the importance of fluid, cross-cutting categorisations rather than opposed, ranked or binary categories
2. great variation in body orientation and skull direction, relating to subtle and cross-cutting age and gender differences rather than major identity differentiation
3. the differentiation of mortuary rites through the introduction of costume, the deposition of individual objects and the scattering of red ochre
4. a tendency for male costume to become increasingly elaborate with age of the deceased
5. a tendency for younger females to have more elaborate costume than children or mature females
6. a tendency for more and more varied grave goods to occur with mature females as compared to younger females and female children
7. an emphasis on the exclusive use of costume elements by adult females or males, counterbalanced by the overlapping use of whole-body costumes for all categories of deceased
8. a tendency for more grave goods categories to co-occur with costumed rather than un-costumed corpses

The analyses of the Kiskőre mortuary domain indicates several cross-cutting trends in the formation of 'personhood'. While most deceased are buried along a specified orientation, the combination of skull and grave orientation produces great variety in all age-sex categories. This is also true for the elements and raw materials comprising mortuary costume – those items in which the body was dressed up. The older the age of death, the more diverse the grave goods accompanying both females and males and the more elaborate the male costume (cf. female costume, which is most elaborate for younger adults). Variations in grave goods and costumes emphasise categorical differences between males and females far more than the differences between adults and children but these differences are expressed not through binary opposition but through the selection of different combinations of elements common to all age-sex classes. It is in this sense that, at the global level, the development of personhood at Kisköre can be described as relational rather than oppositional, based as it is on categories re-stated time and again in the mortuary domain but part of the habitus of the living.

3.4 Micro-tradition analysis

Now that we have defined the global rules pertaining to the total sample of intra-mural burials at Kisköre, it is time to investigate the further potential for uncovering tension at funerals with a micro-tradition analysis of each of the burial lines in the settlement. Seven lines have been identified (*Fig. 1*). The membership of each burial line by age and sex, together with that of the two isolated burials, is presented (Table 5) before a summary of characteristics of each succession of burials by grave line (Table 6).

Table 5. Membership of burial lines by age and sex, Kisköre-Damm.

Group	Females	Males	Children	Unknown	Total
1	23-40* /23-25		2-3	one	4
2	12-14/	46-50/25-27/			3
3	41-50/36-45/				2
4	23-30/23-27/	41-50/41-50/			4
5	23-35/		9-15/10-12/ 9-12/2-5/2-3		6
6	36-45/23-30/23-45/	23-30/ ?one/	one		6
7	50/	58-62/48-52/	12/		4
ISOLATES		25-35	2-4		2
TOTAL	11	9	10	1	31

Note: *(23-40) age range of skeleton as proposed by physical anthropologist.

Table 6. Mortuary treatment of individual burials by order within burial line.

Line/Burial	Grave	Deposited in grave
3/5	74-cm deep rectangular grave	shroud; sherds; 2 red ochre lumps
3/6	99-cm deep rectangular grave	shroud + red ochre; necklace, skin cloak, cattle vertebrae, pot, shell
2/36	65-cm deep irregular pit	shroud; necklace; shell bracelets; armlet; pot; flint blade; shell bead
2/37	106-cm deep grave	bone polisher
2/35	70-cm deep rectangular grave	ochre; 2 shell bead armlets
1/4	45-50-cm deep pit	necklace; armlet; shell bracelet; head-braid; pot; bead necklace
1/1	85-cm deep pit	2 shell bracelets; necklace; pot
1/2	90-cm deep pit	2 necklaces; sherds
1/3	50-cm deep pit	none
7/34	sloping rectangular grave (75-97cm)	2 necklaces; 2 pots; pig femur
7/32	trapezoidal sloping grave (85-130 cm)	red ochre; head-braid
7/28	85-cm deep pit	necklace
7/14	26-cm deep pit (disturbed skeleton)	none
4/7	119-cm deep pit	belt
4/8	top fill of pit	red ochre
4/9	127-cm deep pit	bracelet; necklace; shell bracelet; bead belt
4/21	158-cm deep pit	necklace; head-braid; 2 shell bracelets; chest-braid
6/15	43-cm deep pit	none
6/16	56-cm deep pit (skeleton disturbed)	pottery
6/17	69-cm deep pit	red ochre

6/18	85-cm deep pit	red ochre; shroud + ochre; shells; shell belt
6/19	93-cm deep rectangular grave	red ochre
6/20	115-cm deep rectangular grave	none
5/33	96-cm deep pit	none
5/25	93-cm deep rectangular grave with rounded corners	none
5/31	depth ??, partial skeleton	none
5/26	depth ??	none
5/24	78-cm deep rectangular grave	head-dress; necklace
5/29	60-cm deep pit, damaged skeleton	red ochre
5/30	85-cm deep rectangular grave	none
5/27	85-cm deep pit	none

The Kisköre burial groups consist of between two and six burials, each group with a different set of age-gender categories and with an absence of double burials. The distance of the furthest burial from the centre of the nearest house varies from 7 m to 24 m. With one exception, each group is unambiguously closer to a single house than to any other structure. However, the equidistance of the complex Group 5 between Houses B and C, both of which have their 'own' burial group, suggests a deliberate ambiguity or special status for this line of mainly children's burials.

3.4.1 Burial lines not related to houses

The analysis of the micro-traditions in the Kisköre burial groups emphasises the dialectics of current norms versus new principles. We shall start with groups without houses. Group 3 is the smallest – a dyadic pair comprising two mature females (*Fig. 2*). Their shared age/sex identity is emphasised by the standard burial mode of extended inhumation, standard orientation (SE-NW), a rectangular grave pit and the use of red ochre. However, individual differences are indicated by a different depth of grave and by variations in costume, grave goods and location of ochre. The earlier burial provided the second burial with a set of cultural resources for reinforcement or contradiction.

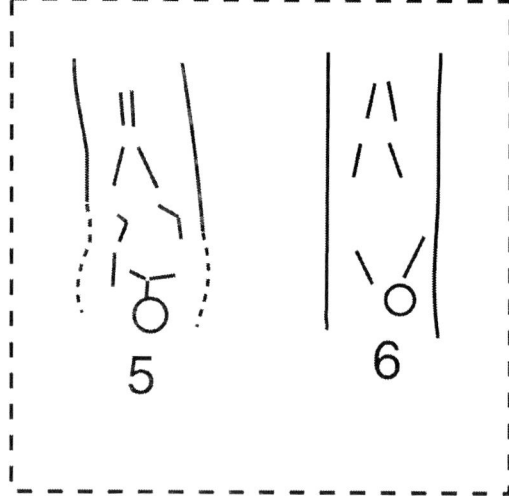

Fig. 2. Group 3, Kisköre-Damm.

In the case of the three burials in Group 2, temporal priority is also unknown but the same result is produced: a strong contrast between the similar outer burials – two adult males - and the middle burial – a female child. These age-sex differences are marked by similarities between the outer burials, and difference from the middle, in the form of the grave (irregular grave/pit/rectangular grave), its orientation (standard on left side/alternative/standard on left side), its depth (shallow/deep/shallow), the use of ochre (absent/present/absent) and the use of costume (elaborate/absent/minimal), in addition to the standard mode of burial for all three graves. However, the differences between the outer graves, relating to age – mature adult vs. young adult – are also significant: a more varied costume and set of grave goods occurs with the older male, who is nonetheless buried in an irregular grave. This is the only series of burials where the maximum number of contrasts – six – is used. The differences continue to a more detailed level, as in the different armlet forms used – limestone beads for the SW burial, shell beads for the NW burial. Here, age-sex principles acting between three closely related persons are used to differentiate each successive burial from the previous one, while a secondary age principle is used for further differentiation. The choice of one of the several alternative body orientations for the female child would appear to relate her burial to the AVK ancestral tradition.

Four burials in Group 1, placed in a line from SW to NE, form the third group without a house: child/young adult female/??/young adult male. This group exhibits two important decisions – group features shared among all burials (extended inhumations in pit graves, with standard orientation in 3 out of 4 graves), and a directionality in one or more mortuary dimensions. Two negative group features are found – the absence of rectangular graves and the absence of red ochre. If the burial line began at the SW end, there was a fall-off in the variety of costume and grave goods (or the converse if the NE burial was the first !):

Table 7. Grave good and costume contrasts in Group 1, Kisköre-Damm.

Grave 4	Grave 1	Grave 2	Grave 3
whole pot	whole pot	sherd	-
headscarf	-	-	-
2 necklaces	necklace	2 necklaces	-
bracelet	2 bracelets	-	-
armlet	-	-	-

The necklaces with three of the burials share the same limestone beads, while the first and third burials share perforated red deer teeth in contrast to the fired clay of the second burial. A summary of differences between successive burials shows low-level differentiation, masked by the missing data on ageing and sexing in the third burial: differences between burials 4 and 1 in grave depth, costume and grave goods; between 1 and 2 in costume and grave goods and between 2 and 3 in grave depth and orientation, costume and grave goods. These variations in burial 3 are combined with the ceramics deposited and the choice of alternative orientation to produce a startlingly different mortuary representation. Is this representation partly prefigured in the directional trends in offerings and costumes or is this the first of the group ? In an attempt to reduce temporal ambiguities, we now turn to burial groups associated with houses.

3.4.2 Burial lines related to houses

Group 7 – a set of four burials grouped around House D – is a classic example of a Household Cluster (*Fig. 3*). However, in the absence of evidence for a house door, it is not clear which burial has priority. One potential temporal clue is the proximity of burial 14 to the AVK burial line but this burial is, in turn, furthest from House D. The initial hypothesis is that the burial nearest to the house is the first interment, leading to the sequence: mature female (burial 34)/ mature male (32)/ female child (28)/ mature male (14). However, since there are no directional trends in burials in this group, this order may perhaps be reversed. In either case, the main differences in this group are between two pairs of burials.

The group is defined by the standard burial mode (extended inhumation) and a special body position – the standard SE-NW orientation, with head facing up. Since only two bodies are placed in this way on the rest of the site, this

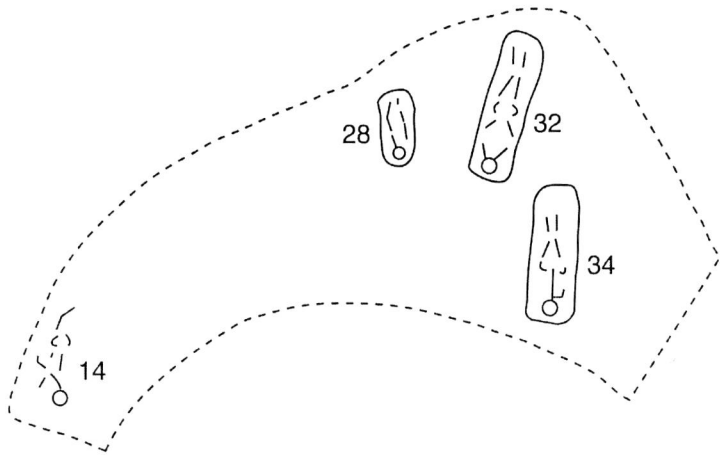

Fig. 3. Group 7, Kisköre-Damm.

position is a strong sign of this group and, by implication, the members of House D. Grave offerings and ochre are deposited sparingly but the main variations in the group are in grave pit depth and form and in body costume:

Table 8. Grave form and costume contrasts in Group 7, Kisköre-Damm.

Grave 34	Grave 32	Grave 28	Grave 14
sloping rectangular grave	sloping trapezoidal grave	-	-
2 necklaces	head ornament + ochre	necklace	-

The unusual sloping graves with their extreme depth (97 cm and 130 cm at the deeper ends) do more to link the mature female and male buried there than to reinforce their sex difference; perhaps a marriage or brother-sister link is documented here ? In any case, the other two burials are differentiated from the first pair by burial in shallower pits – a difference only partly transcended by the similarity in the form of necklaces (limestone beads and perforated red deer teeth) found in the first and third burials. But plotting of successive differences shows that variations decline with distance from the house: between burials 34 and 32, grave depth and form, costume, red ochre and grave goods; between 32 and 28, grave depth and form, costume and red ochre; and between 28 and 14, grave depth and costume. Thus, Group 7 is distinguished both by its body position and by its sloping graves, the latter not found elsewhere at Kisköre.

The sloping graves are used to mark within-group difference as well as between-group differences, while other mortuary dimensions more frequently used in other groups (costume, ochre, offerings) are used less actively here.

Taking the principle that the burial nearest the house is the earliest interment, four burials near House A constitute Group 4 (*Fig. 4*), with a sequence moving further from the house: young adult female (burial 7)/ mature male (8)/ mature male (9)/ young adult female (21). Alongside the standard burial mode and a predominance of standard orientation in 3 out of 4 graves, two directional trends appear to override the two gender pairs in this group: an increase in burial pit depth and greater elaboration of costume with distance from the house:

Table 9. Grave costume contrasts in Group 4, Kisköre-Damm.

Grave 34	Grave 32	Grave 28	Grave 14
-	ochre near head	-	head-dress
-	-	necklace	necklace
-	-	bracelet	bracelets
-	-	-	chest-decoration
belt	-	belt	-

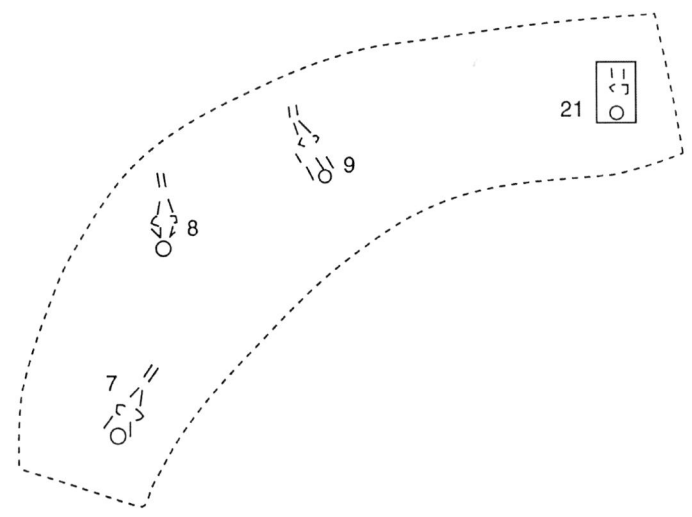

Fig. 4. Group 4, Kisköre-Damm.

Although the costume elements appear to overlap, the raw materials differ in subtle ways (belts of limestone vs. limestone + fired clay; bracelets of limestone beads vs. *Spondylus* shell; necklaces of limestone beads and deer teeth necklaces vs. those elements and perforated *Spondylus* shell). The identity of the same-sex pairs is also overridden by two negative common features – a lack of regular graves and an absence of grave goods, with further contrasts in body position: standard on the right / standard on the right / alternative on the left / standard ?. It is intriguing that, despite the age-sex differences of the two pairs in Group 4, not a single burial or mortuary trait can be unambiguously related to age- or sex-based identity! Rather, this group exemplifies the significance of material culture in creating difference from the community norm as well as intra-group difference from burial to burial through directional variations: differences between burials 7 and 8 – grave depth and form, costume and red ochre; between 8 and 9, grave depth, form and orientation, costume and red ochre, and between 9 and 21, grave depth and orientation and costume.

Perhaps the most regular line of burials is Group 6 (*Fig. 5*), which extends 12 m NE from House C, stopping 6 m short of the AVK ancestral burial line. This line includes 6 burials, in a sequence moving further from the house: young adult male (burial 15)/ young adult female (16)/ young adult female (17)/ child (18)/ adult male (19)/ mature female (20). The group is defined negatively by the near-complete absence of body costume except for one ochre-strewn shroud on the child's burial and, positively, by the standard burial mode and by the near-complete dominance of standard body orientation, once the first burial had been completed. There is no age/sex criterion for the exclusion of costume as a multifaceted, colourful and impressive part of the funeral rites; perhaps the material circumstances of household C or its kinsfolk did not allow for the deposition of such finery. As with Group 4, the main directional feature in

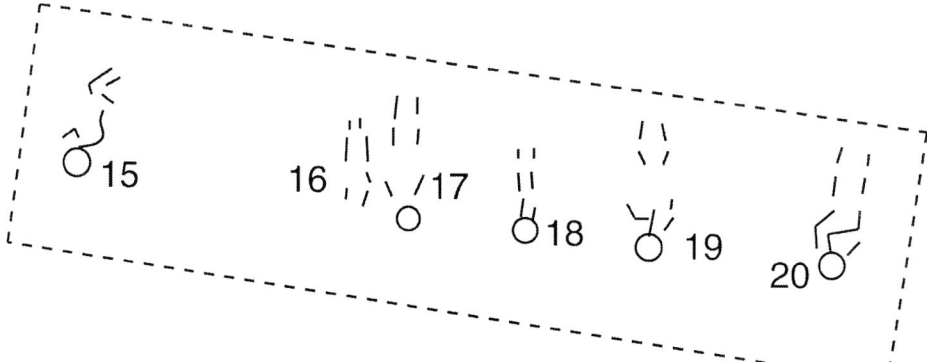

Fig. 5. Group 6, Kisköre-Damm.

Group 6 is the increasing depth of burial pits with distance from the house; it appears that bodily depth is cross-referenced to kinship distance from the ancestral house-builders. The sporadic use of grave offerings and ochre is another means of differentiating successive burials through individual difference: none / pot / ochre / ochre + shell / ochre / none. The final within-group difference is the appearance of rectangular graves for the last two burials, by the same token the deepest graves in the line. This group exhibits no age- or sex-linked material differentiation; rather, the absence of the potential differentiation in costume and the overriding directional increase in burial depth indicates overall homogeneity, with a generally low level of successive individual differentiation: differences between burials 15 and 16, grave depth and orientation and grave goods, between 16 and 17, grave depth, red ochre and grave goods; between 17 and 18, grave depth, costume and grave goods; between 18 and 19, grave depth and form, costume, red ochre and grave goods; and between 19 and 20, grave depth and red ochre.

The final group exhibits by far the most complex spatial structure. Group 5 (*Fig. 6*) may in fact be an amalgam of three sub-groups, 5A – two burials to the NE of the AVK ancestral burial line; 5B – a line of three burials SW of the AVK line; and 5C – a line of three burials parallel to 5B. But there are several important group features which may mean that these eight burials once formed a coherent group. The standard burial mode of extended inhumation pertains; this group contains the highest proportion of children's burials (six out of eight) in the whole settlement; none of the burials uses grave offerings and costume; and ochre appear respectively once. If this coherence is genuine, it seems likely that Group 5 was the key group in the whole burial area, since its structure was based upon the unification of the two settlement zones either side of the AVK ancestral line by transcending that line with a sequence of burials at right angles

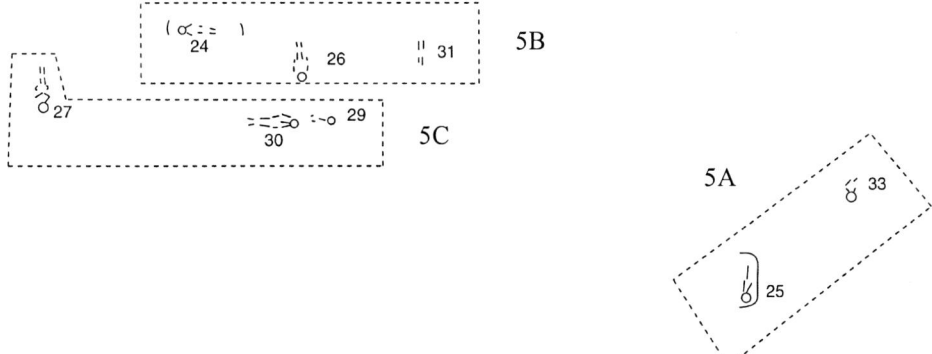

Fig. 6. Group 5, Kisköre-Damm.

to the grave line. However, the possibility remains that sub-group 5A is closely connected to House B, while sub-groups 5B and C, on the other side of the AVK line, are more closely related to House C, if not so closely related to it as Group 6. One repeated means of differentiation was the grave form, with contrasts in each successive pair of graves.

In this analysis, it is assumed that Group 5 forms a coherent group consisting of three sub-groups. Since sub-group 5A is closest to House B, this is taken to be the earliest group of burials. The female child (burial 33) and young adult male (burial 25) were both buried without grave goods, in irregular pits dug to similar depth and according to the standard orientation, although the skulls faced different ways. In sub-group 5B, the three burials /young adult female (burial 31)/ female child (26)/ female youth (24)/ were poorly preserved, but with variable orientation – standard$^?$ / standard$^?$ / alternative on the left. Burial 24 was adorned with a limestone-bead head-dress and necklace. A similar variety in orientation is found in sub-group 5C, with its three female burials – child (burial 29)/ youth (30)/ youth (27): alternative$^?$ / alternative on the left / standard^{UP}. Almost all of the 'alternative' orientations in Group 5 are closely related to the norms of the AVK burial line – the SW-NE direction. The exception is burial 30, with her head to the NE and feet to the SW – a unique occurrence in the Tisza burials at Kisköre but found in AVK burial 11. Thus, in two ways, Tisza-period practice takes a key aspect of the ancestral past and transforms it into the rite of extended inhumation which was in the process of becoming the new Tisza norm. The three instances of the head facing upwards parallel the ^{UP} body position characteristic of burials in nearby Group 7. Hence, materialised references to the past (the AVK line) and coeval neighbours (Group 7) typify the Group 5 burials and support that group's internal homogeneity. The paucity of grave offerings, costume and ochre leave few other means for the individual differentiation of successive burials: differences in grave form between burials 33 and 25; grave form between 25 and 31; none between 31 and 26; grave form and orientation and costume between 26 and 24; grave depth and form, costume and red ochre between 24 and 29; grave depth, form and orientation and red ochre between 29 and 30; and grave form and orientation between 30 and 27. This group includes the only pair of successive burials on site (nos. 31 and 26) which are essentially identical.

3. 4. 3 Personhood and group identities

This detailed examination of a small number of intramural burials at Kisköre-Damm yields a pattern of cross-cutting principles of variability, in which we can begin to see the effects of group identity, age, sex, and individual difference on the material remains. The first question relates to the manner in which group identities were materialised. In Table 10, the features common to all, or all but one of the, members of the group are defined together with any directional traits which define the group trajectory irrespective of age and gender identities. In the Group Features column of this table, a trait listed as 'rare' means that it occurs only once.

Table 10. Group features and directional traits, Kisköre-Damm.

No.	Group Traits	Group Directional Traits
3	mode of burial, grave form, orientation	n/a
2	mode of burial,	n/a
1	mode of burial, orientation, grave pits; no ochre	decreasing elaboration of costume and ceramic grave goods
7	mode of burial, orientation, body position; sloping grave form;	-
4	mode of burial, orientation, no graves; no grave goods; ochre rare; increasing pit; depth; increasing elaboration of costume	-
6	mode of burial, costume rare; mostly standard orientation	increasing pit depth
5	mode of burial, lots of children's burials; no grave goods; costume rare; ochre rare	-

The striking observation about the group features is that all of the six main dimensions of variability are used, either positively or negatively, to proclaim group identity. If group traits and group directional traits are compared with the global rules of mortuary practice, the intriguing thing is that so few of the 'local' characteristics are related to the global patterning. An example is the global tendency for increasingly elaborate costumes to be associated with older males and young adult females. In the two groups where costume elaboration changes directionally, no sex-based relationship is found. Again, while the global pattern suggests a direct relationship between increased frequency of grave goods and more elaborately costumed corpses, the group patterns offer no 'local' support for this proposition.

These findings suggest that the local is not a mere 'reflection' of the global – the communal writ small! Instead, the micro-traditions established over unknown temporal durations in the seven groups at Kisköre indicate considerable variability in their mortuary practices. The internal site chronology is, unfortunately, quite unknown – we cannot be sure whether we are talking about intervals between burials of a month, a year, a decade or a century, though the latter would seem improbable for the given community size. Nonetheless, what we see here is a series of different households, drawing upon varied intra-community and some extra-community relations, making strategic use of material/symbolic resources to construct their own social order. Although there are instances of local 'deviation' from the constructed norm (e.g., the only burial in Group 7 where ochre was scattered), often every burial in the group conformed to the emergent choice of group traits. This shows how individual cumulative actions become the structure which establishes a further cultural context for action. This leads to two questions: how does the 'global' emerge from myriad local versions?; and how does this finding relate to the negotiation of each successive interment against the micro-tradition of earlier burials?

While the 'global' is, in one sense, a post-hoc, etic statistical summary of all the burial acts documented at Kisköre, the existence of these general rules suggests that they were the result of meaningful, emic decision-making. The most likely explanation is that household heads and other important individuals negotiated those aspects of mortuary practice which sustained those community links necessary to an overall group identity. Clearly, such overall bonds remained in tension with other aspects of group and individual identity; those aspects of identity to be drawn upon in any specific burial would most likely have been subject to the tension at funerals which typifies the importance of the mortuary zone.

It is clear from a summary of the group differences in successive burials that the full range of material culture and grave form options was utilised in the pursuit of difference. The extremes of differentiation are relatively rare – with only one burial identical to an adjoining one (Group 5) and only two pairs using the full range of six dimensions to express maximal difference (Group 2). Instances of directional trends are highlighted in Table 10, because they represent the becoming of structure in the micro-traditions of these burial groups. They define the co-emergence of group structure, through the directionality of the trends, and personal identity, through contrasts with each previous burial. The remaining dimensions in groups with such directional trends are statements more concerned with age, sex and individual identities.

In groups with no discernible directional trends, the structure of the micro-tradition emerges in the positive or negative selection of group traits (Table 10). Deviation from these traits marks an emphasis on individual identity rather than group orientation, as in the use of grave goods and red ochre each in a single burial in Group 7. Other aspects of identity are also at work. The oppositions represented in the three burials of Group 2 show the strongest age-sex differentiation in the community, underlining age-sex tensions latent in the community or a strategic emphasis on difference. The variations in grave form, costume and red ochre in the last five burials in sub-groups 5B and 5C – burials of the same age, sex and presumably household group – suggest that individual differentiation between the newly-dead and the most recent burial was often important.

Working through the Kisköre materials enables an understanding of how small-scale actions can form the beginning of a micro-tradition, just as placing two burials side by side can start a burial line. The Kisköre community was so small – probably not more than 50 people at any one time – that it can be assumed that all of the living could have attended the funeral of each newly-dead, even if not all of them did attend. These open and public events were the setting for the crystallisation of action and structure and the material objects were as much part of that setting as the spatial location of the burial. How does the body of ideas about objectification as agency relate to the Kisköre study?

3.4.4 Tension at funerals

Although processualists from Saxe (1970) and Binford (1971) onwards recognised that funerals were a vital moment for community self-reflection and statements about cultural order, they rarely used this insight to deal with small-scale, 'local' processes, preferring to amalgamate data to make 'global' community-wide statements. Agency theory rejects neither the one nor the other approach but seeks to integrate the global and the local by means of an understanding of the material environment in which engendering takes its place.

The final journey of the corpse of each of the newly-dead is, on the one hand, a representation of what the living wish to say about that person, their community and their views of the world – a cultural externalisation. On the other hand, the journey is a movement through and towards sites of consumption, where a final sublation takes place for both the deceased and the living. In this sense, mortuary rites are but one form of social practice rooted in the wider principles of objectification through which individuals and culture are interwoven in the fabric of the quotidian. But what does this mean for the community of mourners who have just lost a two-year-old girl or have to come to terms with the loss of the oldest and (?) most respected member of their community, a sexagenarian female? We begin with the reflexive relationship between people and things.

The two-year old burial 29 (sub-group 5C) is interred next to a 10-12-year-old, in a shallower pit (not a grave), in the same SW-NE orientation (not standard for the site), similarly without grave goods, but with red ochre powdered on the skull (unlike the 10-12-year-old, who had limestone-bead costume elements on head and neck). Here is a formal burial of a very young member of a community, whose journey of self-definition had hardly begun. Yet the household defined that child's persona in a formal way, using material presences and absences to make references to wider cultural themes (the presence of ochre; the absence of grave goods) and linking the girl's death to the AVK ancestral line through body orientation. It is not known whether the child burial 29 was the first such to be buried at Kisköre but the formal burial itself shows that the category of 'two-year-old girl' was socially recognised as emergent at the time of the child's funeral. The funeral, as a public act of mediation between the mourners and a wider society, provided the elements of cultural identity for the newly-dead. The precise form of burial which we see is, in part, a snapshot of the social processes of categorisation, the practice of description of a particular individual who, at the same time, is a member of a wider category of people. If we do not know the ages and sexes of the people who made the representation, we are nonetheless aware of the wider mortuary discourse used for sex-based categorisation. But other factors intervene in choice of mortuary practice – the time of year of the burial, the quality of the harvest, the state of inter-group relations, etc, etc. We can discern only some of these elements which made each burial a profoundly human, as well as cultural, act.

In contrast to the oldest member of the community, few material objects were laid to rest with the two-year-old burial 29. If, as we suppose, the mature female burial 34 is likely to have been the first burial made in Group 7, her funeral set the direction of the group's mortuary practices. Her sloping rect-angular grave form was unique to the site, being imitated in a variant form only once, in the immediately following burial. Her body position (SE-NW^{UP}) initiated what became a defining group feature but was rare outside the group. Her costume was cross-referenced to other burials and social categories in two ways. The use of necklaces without other costume elements is found with other women and with children. The precise form of necklace – made of a com-bination of limestone beads and perforated red deer teeth – is the commonest form of necklace, found in all age-sex categories on the site and in burials in four Groups; both elements of the necklace make reference to wider social networks. In terms of grave goods, the only objects deposited in Group 7 burials were those in burial 34. The deposition of fragments of two hemispherical Tisza bowls on the right side of her body and a pig femur near her left knee creates an opposition between nature/left and culture/right that is rarely replicated at Kisköre; both of the bowls were large enough and ornate enough to have been used as serving dishes for the joint of pork, an offering unique at Kisköre. Hemispherical bowls were placed in only two other burials, both mature individuals, one female (burial 6) and one male (burial 36), and in both instances on the right side of the body. Through the categorisation of people and things, objects such as the necklaces, pottery and pork joint of burial 34 integrate the person within the normative order of the social group, not only at household level, but also in the wider community and beyond. Whoever made the decisions leading to the particular cultural form of burial 34, they were adhering to a repertoire of material forms and social practices whose repetition, with variations, created long-term structure. Their choice to associate the female in her sixties with specific costume elements and objects came out of past engendered practices and led to their persistence.

The second structuring relationship concerns the relationship between people and places, as betokened in the spatial relations of the site. Kisköre shows, in microcosm, the closeness of household life and death, the familiarity of nearby places and the tight nexus of domestic and mortuary arenas, not yet transcended until the succeeding Copper Age (Chapman 1994, 1997).

A primary factor is the location of the AVK ancestral line bisecting the site. The burial groups are as strongly attracted by the magnet of the AVC line as are the houses, all of whose nearest walls lie within 30 m of a point on the AVK line. This line grounds the Late Neolithic settlement in a past, timeless tradition, while at the same time providing cultural material on which to draw for similarities, contrasts and oppositions. After the burial group is located in relation to the ancestral line, the choice of direction of the burial line is important. It is perhaps significant that burial lines 4-7 all expand in the direction of the AVK ancestral line or its extension; in the case of Group 5, the AVK line is crossed while, in the case of Group 4, the line stops short of an extension of the AVK line. This double influence of the ancestral line, on the origin and the end-point of several lines, says much for the overall spatial structure of the site in relation to temporal principles.

It is also clear by now that the location of successive burials in a straight or curved line increasing in distance from a house is another form of self-structuring social practice. A linear arrangement of burials provides a kinship calculus defined not only by inclusion / exclusion but also by relative position, particularly distance from the house and distance from the ancestral line. Because most burials lie between their households and the ancestral AVK line, there is movement along the burial line between the living and the ancestors, with the earliest burials closest to the living and the latest nearer to the long-dead. This unilinear spatial principle cuts across the cyclical movement of 'individuals' between unborn, living, newly-dead and ancestor and grounds the corporate group in a unique micro-traditional history. It also expresses the tension between the living and the dead in a way not always foregrounded in spatially distinct cemeteries.

Another general spatial structuring principle concerns house and burial orientation, since the latter could have drawn upon the cultural resources of the former. This in fact happens relatively rarely. Only two out of the six houses (Houses A and E) were oriented in the standard SE-NW burial line and one of these (House E) had no associated burials. In three cases, all the burials in a group were placed on different orientations from that of their associated house (Group 7 and House D; Group 6 and House C; Group 5 and House B). The other house with no associated burials, House F, is oriented SW-NE, as is House C and several burials from Group 5B-C. At this site, the avoidance of house orientation is an important structuring principle of burials, as if to emphasise the difference between the living and the dead. This is not the case in other parts of Europe, for instance the Neolithic of the north European plain, where long mounds often mimic the orientation of long houses (Hodder 1984).

A third aspect of structuring in the Kisköre burials is the manner in which the community related its social practices to other Tisza settlements and their social practices. In the most recent general summary of Tisza and other Late Neolithic mortuary practices in the Alföld Plain, Kalicz and Raczky (1987) outline the salient practices of tell burial: small groups of pit burials, occasionally in coffins, common orientation of adult males, often in contrast to the orientation of adult females, frequent use of red ochre and restrained use of grave goods as a form of differentiation (see also other chapters in Raczky 1987). This summary indicates that the main forms of variability were common throughout the Late Neolithic of eastern Hungary but that different communities drew differently upon these traditions to tell often similar stories about their community's newly-dead. An example is the use of the predominant NW-SE orientation – rather than the SE-NW orientation at Kisköre – of the burials at the low tell of Öcsöd, with adult males lying on the right side and adult females lying on the left side (Raczky 1987a). This picture is remarkably similar to the pattern in which three Bulgarian Late Copper Age communities made different, village-specific use of broadly similar material culture to send messages about age and gender categorisation in eastern Bulgaria (Chapman 1996). We may suppose that, had a Tisza-group family from another settlement been present at a burial in Kisköre village, they would have been familiar with the overall practices but perhaps mystified at the differences in detail from their own traditions, prompting them to wonder why 'those Kisköre people' were a little bit strange!

In short, there were four ways in which structuring was built into the Kisköre burials: the accumulated set of associations between persons and objects; the spatial relations which developed with the slowly growing number of burials; the embedding of this particular community's mortuary practices in a wider discourse of regional trends; and the presencing of the extensive set of exchange relationships betokened by the material culture of the Kisköre grave goods. At any time, the people gathered around the burial site of a newly-deceased member of the community had many choices for the exact form of burial. Two choices in particular weighed on those mourners: the choice of how differently to bury their kith and kin from the last burial in their burial group and the choice of how to relate the ceremony to other burials within the village. The tension between the household's micro-tradition and the potential inherent in the new statement shortly to be made about the newly-dead encapsulates the dialectic of structure and agency within an enfolding debate about self-identity at both individual and community levels.

3.5 Summary

In this chapter, I have defined the tension at funerals as a microcosm of the structure – agency dialectic. This dialectic is encompassed within a broader process of cultural constitution – the objectification of people through their externalisation in things and places and the subsequent sublation of the values and identities created through materialisation. I propose that the process of objectification is the principal form of structuring through which people create their own material environments but that people always oppose past traditions – the naturalisation of material forms – through cultural resistance. As far as agency theory is concerned, the key feature of objectification is that it is an individual process through which group norms and past traditions are constraining but not determining. The representation of humans as cultural categories is the central material form of individual externalisation and its medium is a very political choice. Sex-based identities are created through the process termed 'dynamic nominalism', in which new categories of people come into existence at the same time as the people who fill those categories.

This works as much in the mortuary domain as in any other cultural context, through the form of the burial rituals selected for any given individual.

It is the sequence of burials and their spatial locations within a place already heavy with the past which lends salience to the set of intramural burials from the Hungarian Late Neolithic site of Kisköre-Damm. Most of the 31 burials are grouped into seven burial sets, including arcs and lines. Most of the burial lines are associated with a specific house although some lines have no house and two houses are far from the nearest burial line. Each burial line is treated as a distinct micro-tradition within the totality of community burials and, where it is possible to establish the probable sequence of burials, an analysis is made of the differentiation of each successive burial from the previous one. In this way, it is possible to define 'global' or community traditions, 'local' or micro-traditions and the resistances which individuals make to micro-traditions and local traditions make to the community level.

Three ways of structuring the mortuary domain at Kisköre are defined: associations between persons and grave goods; gradually enfolding spatial relations; and the embedding of community mortuary practices in the regional patterns. This leads to a further tension – between each village and the regional totality of mortuary practices. The negotiation of such tensions at a multiplicity of levels constitutes the framework for the creation of both personal and group identities.

4. The earliest formal cemeteries in Hungary

4.1 Introduction

The first well-documented use of cemeteries as a spatially distinct mortuary arena occurs during the east Hungarian Copper Age. At this time, tell occupation becomes quite rare, although burial still continues on tells. Instead, the dispersed farmstead constitutes the commonest settlement unit (Bognár-Kutzián 1972; Sherratt 1982a, 1982b, 1983). The shift to farmsteads away from tells has three implications for our understanding of social reproduction: (1) the dominance of the household as the primary economic unit, in contrast to the densely packed overlapping social networks on the tells; (2) the importance of extensive local networks linking perhaps as many as 100 dispersed farmsteads into an exogamous breeding unit; and (3) the predominance of a competitive ideology of prestige goods accumulation over the overtly traditional, egalitarian values of the tell village. This third point is based upon the wide range of rich grave goods in the inhumation graves of complete, articulated skeletons that characterise the mortuary population of Early Copper Age cemeteries such as Tiszapolgár-Basatanya (Bognár-Kutzián 1963). These cemeteries are often partitioned into lines or rows of graves, which may represent lineages or household groups from particular farms. Barrett (1990) has shown the significance of placement for individual burials in barrow cemeteries of the Bronze Age of southern Britain (cf. also Mizoguchi 1993). The Copper Age cemeteries show the emergence of complex cemetery topography without a monumental burial form. The denial of monumentality is a form of symbolic distancing from the ancestral tell monuments which stood unoccupied, if not unused, in the landscape.

The Copper Age cemeteries tend to be located some distance not only from tells but also from contemporary farms, giving them a liminal status between the farmland of the living and the world of the ancestors – the locus of transition of states of being. The removal of the newly dead from direct association with the houses of the ancestors (on the tells) and those of the living (the dispersed farms) suggests a different conception of ancestral landscape, more in harmony with the dispersed social relations of an exogamous network than the place-based values of the tell communities. The spatial linearity of the cemeteries is matched by their linear concept of time, with the once-and-for-all insertion of a sequence of dead bodies until dissolution of the lineage leads to abandonment of the cemetery. I now turn to a more detailed analysis of the Basatanya cemetery near Polgár, since it spans the Early-Middle Copper Age and is the largest excavated sample of Copper Age graves in eastern Hungary.

4.2 The Basatanya cemetery: background and previous analyses

The Basatanya cemetery was partially excavated in 1929 by F. Tompa before the main excavation in 1950-54 by I. Bognár-Kutzián as part of her wider project in North East Hungary (Bognár-Kutzián 1963). The site was occupied in the Neolithic period for an unknown duration; remains of Szilmeg-type pottery were found in a number of pits (*Fig. 7*). One Late Neolithic grave has also been recognised. A total of 154 Copper Age graves was identified, 60 from the Early Copper Age (Period I), 87 from the Middle Copper Age (Period II) and seven from the Transitional (I-II) Period (*Fig. 8*). The total duration of the cemetery was estimated at cca. 220 years, at a rate of one burial per annum, by the excavator (Bognár-Kutzián 1963: 351-2) but recent radiocarbon dates on human bone support a much longer period of use, albeit many of the dates have standard deviations of up to 200 years (Benkő et al. 1987: 1000). A series of thermoluminescence dates on ceramics confirms the general date range but the standard deviations of 500-800 years preclude precise dating. On the basis of the new dates, Forenbaher (1993) proposes a duration of 900 years, from 4500 to 3600 CAL BC, implying a mean rate of one burial every six years. It will be important to evaluate whether Bognár-Kutzián's model or that of Forenbaher is more acceptable for the overall development of the cemetery.

Despite field survey around the cemetery, Bognár-Kutzián was unable to locate any coeval settlements. However, in the recent Upper Tisza Project, small scatters of Copper Age pottery were located in fieldwalking South of Bosnjak domb (Chapman – Laszlovszky 1992). These scatters, located less than one km from the cemetery, indicate the kind of small, probably short-term settlements which have been classed as dispersed farmsteads elsewhere in the Plain (Patay 1974). It is thought likely that many of these farmsteads buried their dead in the Basatanya cemetery.

The sequence of burials at Basatanya has been the subject of debate ever since Bognár-Kutzián identified rows of graves in both Periods. While the existence of lines of burials is widely accepted (Skomal 1980; Meisenheimer 1989; Wyatt 1994; Chapman 1997; Sofaer Derevenski 1997, 2000), there has been uncertainty over the placing of individual graves in rows and, on a larger scale, over whether burial in one line is completed before the next row is begun or, instead, whether several rows, perhaps relating to different social units, are still accepting burials at the same time. The excavator favours the former view, based partly on the location of the rows but also on her pottery typology which supports an overlapping distribution of types and sub-types which evolves with time from South to North rather than West to East (Bognár-Kutzián 1963: 229-

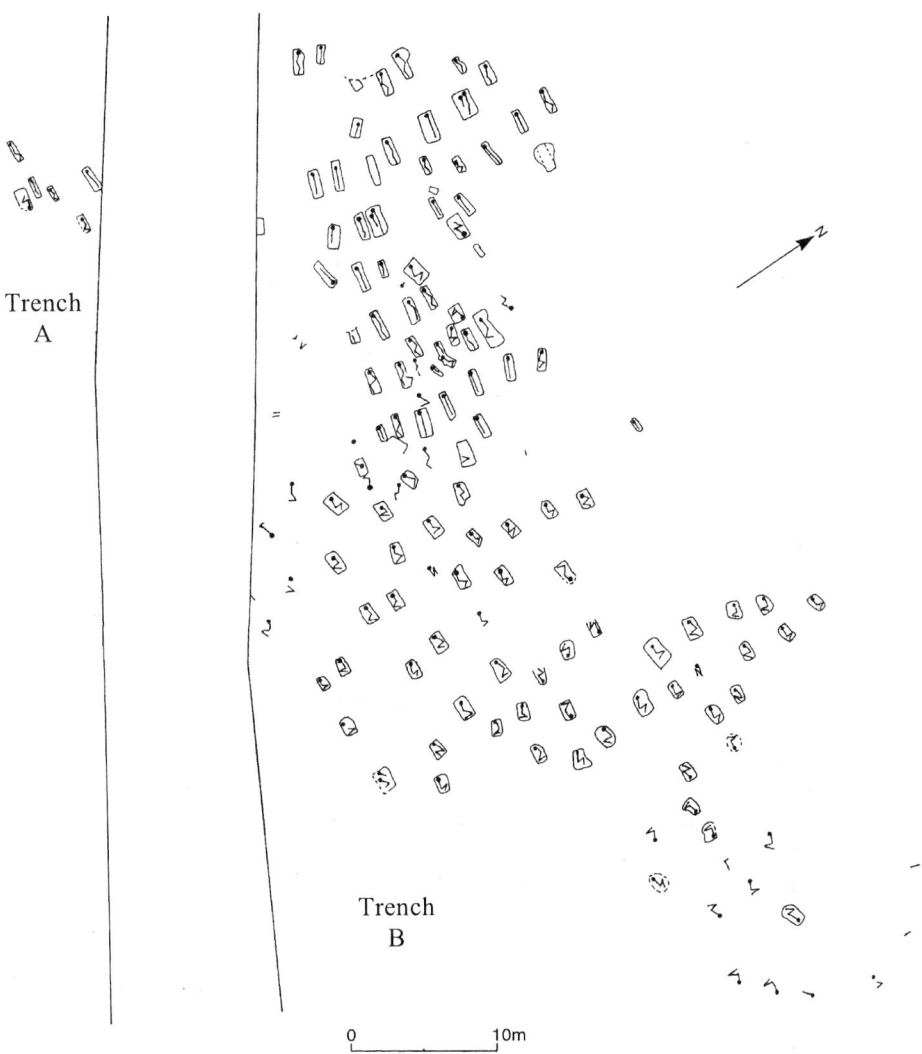

Fig. 7. Simplified plan of Tiszapolgár-Basatanya cemetery.

231, 295-7). This view is also accepted by Wyatt (1994), some of whose analyses are based on the assumption that burials can be placed in an absolute order from first to last.

However, it is important to differentiate two separate but interlinked processes of growth in the cemetery layout: (1) the laying out of individual grave lines in a general South-North direction; (2) the succession of grave lines, which, on the excavator's ceramic typology, shows a directional trend from

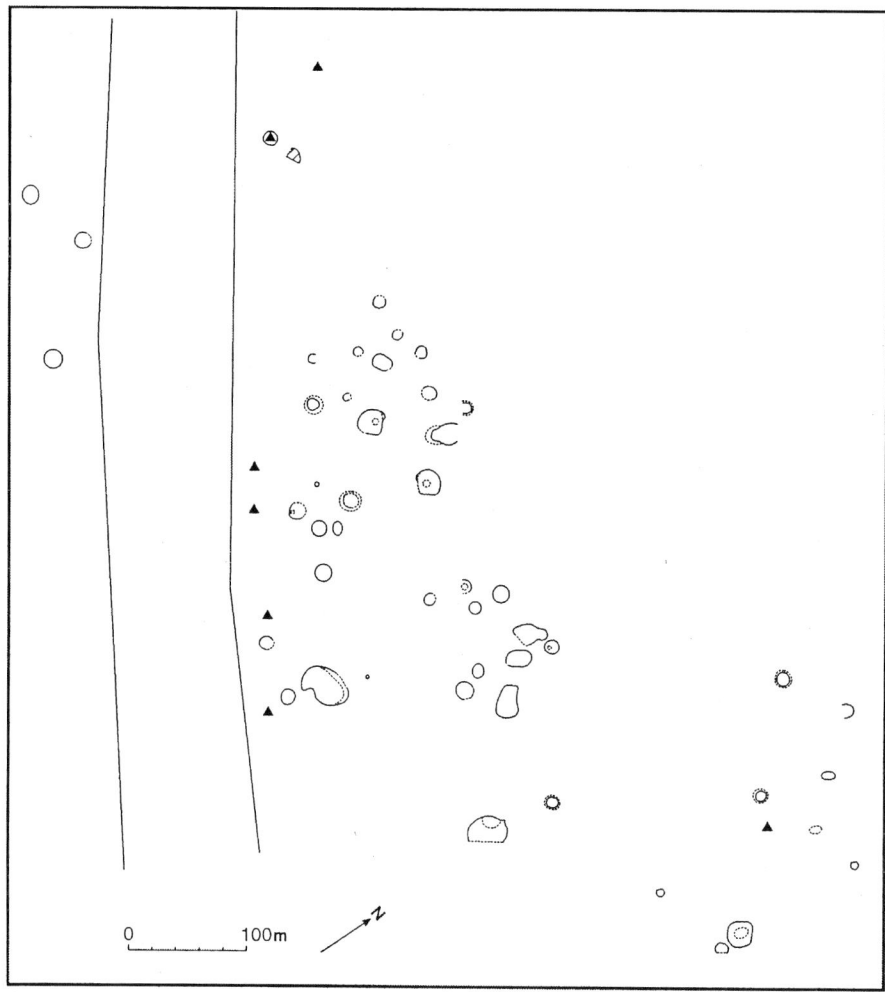

Fig. 8. Plan of Neolithic features at Tiszapolgár-Basatanya in relation to area of burial in Periods I and II.

West to East. This is confirmed by the spatial observation that the grave lines of Period I extend over cca. 45% of the South-North cemetery axis, with Period II grave lines overlapping Period I lines by cca. 15% and taking up two-thirds of the axis. We shall return to the point of the overlap of Period I and II grave lines. For the moment, it is important to note the predominantly West-East development of grave lines. However, within the grave lines, the view that several rows are accepting burials at the same time finds some support in the new radiocarbon dates. Although these dates are not high-precision dates, the dating of graves in the same rows seems sufficiently far apart, at the 68% confidence limit, to suggest a long duration of burials in several rows. An example is what I have termed Line B (see below, pp. 88-90), where dates from graves 23 and 28 fall into the early 4th Millennium CAL BC while the date from the adjacent grave 12 falls into the fifth. This would mean that the burials did not fall into a continuous sequence of successive lines but, rather, different social units (perhaps each dispersed farmstead?) would open up a line and continue to bury in it until they ceased the practice. This alternative gives more analytical prominence to the grave lines, which hitherto have been merely identified and then largely ignored, without denying the overall South-North spread of burials. This suggestion also eliminates the 'problems' caused by the lack of independent space for the Transitional Period I-II graves and the insertion of Period II graves into supposedly long-abandoned Period I grave lines. It may also help to explain the characteristics of individual grave locations in relation to other burials and maybe even other details. It is important to underline that, at Basatanya, as in other large cemeteries, there is a close and recursive relationship between time and place.

Unlike the Kisköre Late Neolithic mortuary data set, many aspects of the Basatanya cemetery have been subject to detailed investigation by the excavator, in addition to partial analyses by later researchers. Bognár-Kutzián's approach is, in some ways, more post-processual than that of later researchers, when she remarks (1963: 10) that: "graves do not reflect life but only the prevailing rites" (cf. Meisenheimer 1989: 2). On the other hand, her evaluation of multiple explanations for the complex data at hand are much more processual in tone, as is her view that "grave goods reflect an objective reality" (1963: 10). This view is matched by her use of an inter-disciplinary approach to gain insights from zoological, technological, stylistic, ethnological, physical anthropological, chemical and pathological analyses to understand prehistoric mortuary practices. Many of the basic rules governing burials at Basatanya were first identified by Bognár-Kutzián.

Skomal (1980) approaches the Basatanya cemetery from the standpoint of historical changes involving the arrival of the Indo-Europeans in Central Europe. This narrative-driven approach relies upon the successful identification of Indo-European type-fossils and assumes that any changes found across the critical Early-Middle Copper Age threshold are, somehow, related to this alleged historical event. A second problem is that Skomal's main methodology requires the lumping of many Copper Age cemetery data sets to provide a gross contrast between the ECA and the MCA – thereby losing any means of recognising that communities drew upon the same material culture in very different ways to establish their identities (e.g., for Bulgarian Copper Age cemeteries: Chapman 1996). Weisenheimer (1989) uses a series of detailed distributions of material culture, especially pottery types and sub-types, to indicate the overlapping development of types through the use of the cemetery, a conclusion not necessarily supporting an easy division of graves into the two main periods. Her main concern is the definition of mortuary rituals, which act as a filter between the dead and the living.

The more recent re-analyses of the Basatanya data have all built upon the excavator's excellent publication to develop insights about changes in the expression of age, gender and identity in the two periods of cemetery usage. Preferring the short chronology for the cemetery and assuming a continuous stream of burials, Wyatt (1994) defines a cyclical distribution of wealthy graves appearing once per generation of 27 years in Period I, a cyclical pattern which is absent in Period II. He argues that the strict rules regarding burial and grave goods deposition in Period I imply the burial of only a sub-set of the total population, while the MCA deceased represent the whole population which, for the first time, enjoys equal access to burial. Sofaer Derevenski (1997) begins with the problem of how to separate social processes of engendering from the biological identification of sex in burials. Extrapolation of the rules for right-sided burial of most adult males and left-side burial for most adult females to the children whose remains could hitherto not be sexed enables her to develop a perspective on the engendering of individuals with increasing age. Sofaer Derevenski distinguished between discrete variations (especially the presence/absence of specific grave goods), which she believes to indicate sexual difference, from continuous variations (e.g., the number of objects in graves), which she takes to mean a learnt, gendered difference. In this approach, the main difference between Periods I and II at Basatanya is the importance of gendered distinctions in the former, through the use of pottery, costume and lithics, and the greater emphasis on sexual differences in Period II through categorical oppositions in material culture.

My own categorical analysis of the two Periods at Basatanya (Chapman 1997) attempts to identify the kinds of grave good categories which are uniquely associated with a specific age/sex category (e.g., only adult females). This study takes into account changing sex ratios (the sex ratio is more or less equal in Period II but males outnumber females in Period I by almost 2:1) and the under-representation of children in both Periods (Nemeskéri – Szathmári 1987). The non-ceramic grave goods fall into 39 types, 28 of which appear in Period I and 29 in Period II. In Period I, the largest number of associations defines the identities of males and children: 15/28 (54%) of categories are associated with these two classes. 7/28 types (25%) occur with all age-sex classes, while females and females with children have few associations.

In Period II, a very different picture emerges: gendered adult identities rather than those of children are reinforced by means of artefact associations (12/29 artefact types for women only plus 10/29 for men only; or 76% combined). What is even more striking is the change in the choice of grave goods for negotiating social identities between Periods I and II. Only 4/28 types used in Period I maintain the same age-sex class association in Period II. While 11/29 types found in Period II are new, fully half of the Period I symbols are associated with new age-sex classes, mostly for adult women and men. The main change concerns the switch from joints of meat, with a symbolic emphasis on hunting, to productive artifacts, such as spindle whorls, metal ingots and flint cores, or status artifacts, such as stone hammer-axes and metalwork, often now found to indicate femaleness. While the associations with hunting do not disappear in Period II, their decline suggests that other activities are perhaps beginning to gain equal status (Chapman 1997). The analysis suggests that the Basatanya cemetery spans a major transformation in the negotiation of gendered social relationships, part of which may be captured through a diachronic analysis of this site.

All of the analyses so far discussed share the same two units of analysis – the individual graves and the Periods at Basatanya (at other 1-period cemeteries, the second unit is the whole site: e.g., Zengővárkony: Dombay 1960). On these criteria, the size of the Basatanya cemetery is one of the principal reasons for its frequent re-analysis, since sample size is important for production of statistically significant results (e.g., Sofaer Derevenski 1997). But the collapsing of all Period I graves into a single unit of general analysis diminishes the possibilities of a more nuanced approach and ignores the mass of spatial data latent in the complex distribution of graves. In particular, since the spatial distribution of the graves is subject to such a strong structure, it would seem highly probable that the social meaning of such structuring would be of importance to an understanding of the cemetery. As emphasised above (pp. 66-68), agency can be explored more successfully at the 'local' level, in tension with the more 'global' structures.

4.3 Global rules at Basatanya

It is useful to summarise the general rules characterising the burial sequence at Basatanya, as derived from Bognár-Kutzián's wide-ranging analyses and the contributions of later researchers. This set of rules permits comparison of the mortuary practices of any of the local groups with the long-term, "standard" structures of the whole mortuary community. These rules may be summarised under 17 headings:

1. The majority of burials is laid out in linear groups, with other burials in groups and a small number as isolates.
2. The trend of the graves in each grave lines is, with a few exceptions, from South to North. However, the layout of the grave lines follows a West-East sequence.
3. Burials were usually made in (irregular) rectangular pits dug to a depth of between 30 and 90 cm.
4. Graves and bodies were usually oriented in a West-East direction, with heads to the West, especially in Period I but with more variation in Period II.
5. The vast majority of burials were individual inhumations, with only a few multiple graves. There is a chronological difference between the flexed body position of Period I and the more tightly contracted body position of Period II, with Transitional I-II graves intermediate.
6. Sexual difference was emphasised by the sidedness of the body, especially in Period I: females were laid out on the left side, males on the right.
7. The adult sex ratio varies with time: almost 2:1 for males:females in Period I and approximately equal in Period II. Children are under-represented in both Periods but especially in Period II, when more middle-aged and older people were buried.
8. The children's burial rite is broadly similar to that of the adults.
9. Individuals showing signs of bone pathology (violent deaths or diseases) are buried in a broadly similar way to 'healthy' people.
10. Grave goods are far commoner in Period I than in Period II.
11. There is an overlapping distribution of pottery types and sub-types, supporting the South-North expansion of the cemetery and continuity in material culture. However, there are differences between Period I and Period II ceramics at the type and, especially, at the sub-type level.
12. Pottery is used as a continuous variable indicating gender in Period I, where the number of pots deposited is age-related. In Period II, there is more discrete variation in pottery deposition, indicating sexual difference.
13. Animal bones are deposited as pets, as food remains or as trophies. Pig mandibles are particularly important in Period I, while the incidence of domestic animal bones increases sharply in Period II.
14. The distinctiveness of female graves is emphasised by the use of limestone bead girdles as a costume element. Variations in male costume is age-related in Period I, but male costume is very rare in Period II.
15. Sexual difference is marked by the association of unworked stone pebbles with females and worked stone with males. There is an age-related distribution of stone blades of differing sizes in male graves.
16. Artifact categories associated with solely adult males or only children are common in Period I, while few categories are associated solely with adult

females. In Period II, far more artifact categories are associated solely with adults, both females and males, than with children.
17. Few artifact categories are associated with the same age/sex categories in both Periods I and II.

These rules are a combination of discrete and continuous differences and provide a summary of those patterns perceived to characterise the whole of the cemetery or either of the two main Periods. They constitute a generalisation of the results of all the analyses, which could be used to compare and contrast Basatanya with all other Copper Age cemeteries, within and outside Hungary. If there is a sense in which a historical narrative can be formed from the results of archaeological research, it is these rules which would typically be taken to create such a story.

However, the tension between structure and agency at the general level operates at the level of mortuary practice through the individual burials positioned not only in relation to the general rules outlined above but also in relation to the immediate social and spatial context of previous interments. While each successive burial partakes of the place-value of the whole cemetery, the immediate antecedent burial and the local grave line or group creates a more intimate context for the newly-dead, in which isolated burials do not share. The tension at funerals which confront mourners is a tension between the local statements which they may wish to make, whether in agreement with or in contrast to the mortuary ritual of an antecedent burial, and the distance they are able to put between that burial and the age-, sex- and social group-related rules pertaining to the cemetery community as a whole.

4.4 The micro-tradition analysis

To begin at the beginning is often difficult at mortuary sites and Basatanya is no exception. The cutting of the Szandalik canal, which exposed and destroyed Copper Age graves, removed a block of terrain linking what the excavator claims to be the earliest burial on site to the main area of the Copper Age cemetery (or Trench A) (Bognár-Kutzián 1963: 158, 518 – Supp. 1: here = *Fig. 7*). In this isolated excavation trench B, a group of five Period I graves is found adjacent to grave 84, a strongly contracted inhumation oriented East-West, with a Csőszhalom-type tumbler as a grave good. These characteristics support a Late Neolithic date for this apparently isolated burial – the earliest known on site. Unfortunately, the Szandalik canal has removed any evidence for the linking of the Period I burials closest to grave 84 to the Period I graves in Trench A but it is likely on the grounds of location and orientation that the

Trench B Period I graves would have been connected to a very early grave line in Trench A.

It may be hypothesised that the Late Neolithic group responsible for the burial were conscious of the Middle Neolithic Szilmeg domestic occupation and used it to relate their newly-dead to an ancestral realm, as at Kisköre (see Chapter 3). A second stage of ancestral links was then provided by the location of the Early Copper Age cemetery close to the Late Neolithic burial.

The absence of any other Neolithic burial in the excavated areas which, otherwise, contained almost 40 Neolithic pits (*Fig. 8*), is a significant negative. It would appear to indicate that Late Neolithic grave 84 was used as a foundation deposit for the cemetery, a starting-point which linked the earliest Copper Age burials to an ancestral presence. This may well signify that the ancestral location is to the South, the direction at which Copper Age burial lines begin. We cannot be certain that a South-North sequence occurred in every grave line but the likelihood is that the location of the ancestors is a significant part of community tradition and myth. For the purposes of the micro-tradition analysis, the assumption is that they ran from South to North.

4.4.1 The Period I-II transition at Basatanya

Before we can proceed to an analysis of the grave lines of Period I, it is important to clarify the position of the centre of the cemetery concerning the spatial distribution of graves from Periods I, II and the I-II transition (*Fig. 9*). This distribution indicates a complex process of interlocking graves. Bognár-Kutzián confronted the problem in two ways: (1) the inference of a sequence of graves (1963: 229-231) and (2) a discussion of reasons for the location of Period II graves in a part of the cemetery dominated by Period I graves (1963: 519-520). In this latter section, the excavator ignored the question of the Period I-II graves, the majority of which are also found in the 'Period I' part of the cemetery. There is also the question of the two graves dated to Period I which are found in the 'Period II' part of the cemetery (graves 81 and 80), as well as a third grave (82 – dated to Period I) found as a remote outlier but closest to a line of Period II graves.

One approach to these questions of temporal and spatial order was the typological and distributional study by Meisenheimer (1989), who has mapped the locations of all her ceramic types and sub-types, as well as many other aspects of mortuary practices. Several conclusions can be drawn from her work. The main point is that, although there are some ceramic types which allow a good differentiation of Periods I and II (notably changes in bowl types and the replacement of amphorae by milk jugs in Period II: 1989: Karte 3), the distribution of the vast majority of ceramic types and sub-types shows significant overlaps between Periods I and II (1989: Karten 5-18, 20-31). This is confirmed on the two ceramic matrices for Periods I and II (1989: Tabelle 1a-2) and well illustrated by Meisenheimer's map of pottery said to resemble Period I types which occur in Period II graves, a practice which continued well into the latest grave lines (1989: Karte 32). The other most convincing differentiation between the two Periods is on the basis of the body position, with a significantly greater degree of contraction in Period II graves (1989: Karte 34). The end-result is that, while the extremes of Periods I and II can be clearly distinguished one from another, it is increasingly difficult to divide those graves in the latest phase of Period I from those earliest graves in Period II. The principle proposed here is that, provided the grave is in a close relationship to other graves in a potential group, its dating to the I-II transition or even to Period II will not prevent a certain grave from being considered in relationship to the other graves of Period I.

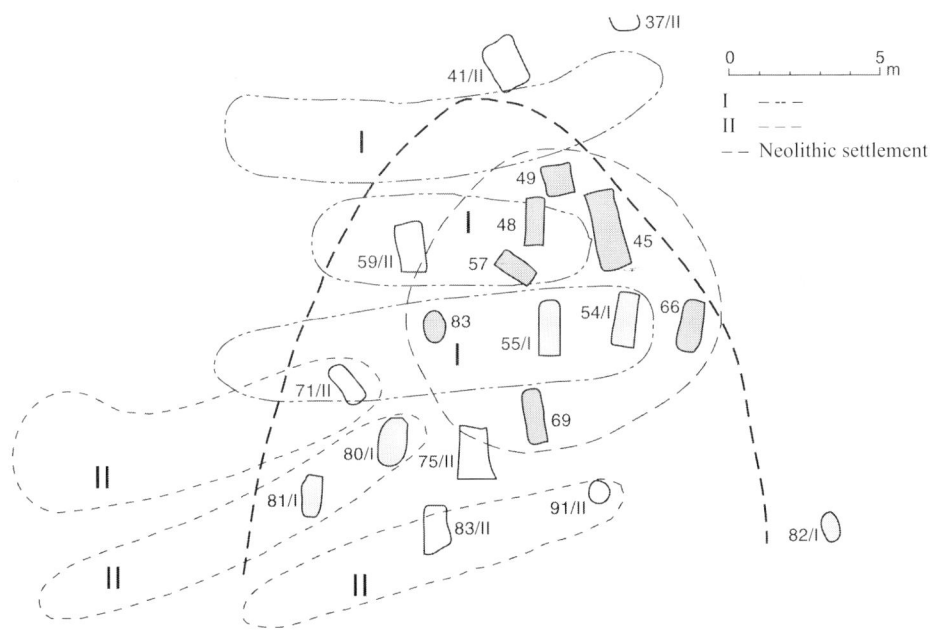

Fig. 9. Plan of transitional Period I-II graves
with adjacent Period I and II graves, Tiszapolgár-Basatanya.

4.4.2 Period I grave lines

Eight grave lines can be recognised in the Period I part of the cemetery. All of these lines were recognised indirectly by the excavator in her sequence of burials (Bognár-Kutzián 1963: 229-231) but there have been minor modifications relating to the three criteria for inclusion in a grave line: (1) there should be a minimum of three graves in a line; (2) the graves should share the same, or a similar, alignment; and (3) the distance between graves should not exceed 4m or double the length of the average grave. Thus, grave 77 is excluded from the line containing graves 27-29, 23, 12-13 and 10 because its alignment is different and it lies over 15m from grave 27. This leaves a total of 13 graves not in lines – two groups (76-79 + 88; 80-81) and six isolated graves (9, 18, 19, 22, 34 and 82). The age-sex composition of the grave lines and the other graves is as follows (Table 11):

Table 11. Age/sex composition of grave lines and other burials, Period I, Tiszapolgár-Basatanya.

Group	Adult Female	Adult Male	Female Child	Male Child	Incomplete Information	Cenotaph	Total (Path)
A	1	1	2	-	-	1	5 (2)
B	1	5	-	1	-	1	8 (4)
C	1	2	2	-	-	-	5 (2)
D	3	2	1	-	2	-	8 (3)
E	-	2	-	1	-	-	3 (1)
F	-	4	-	-	1	-	5 (2)
G	3	6	-	2	-	-	11 (4)
H	4	2	-	2	1	-	7 (4)
Sub-T	13	24	5	6	4	2	54 (22)
Other	3	2	1	4	4	-	14 (5)
TOTAL	16	26	6	10	8	2	68 (27)

* Note: two Period I burials are double burials, yielding a total of 68 bodies from 66 graves.
Key: (path) = number of skeletons with a pathological condition.

It is interesting to note that, despite under-representation of children, at least one burial of an adult female, an adult male and a child occurred in six out of the eight grave lines. This does not contradict the notion that the grave lines were, for the most part, family lines, perhaps deriving from different dispersed settlements. In addition, every single grave line contains one or more burials with pathologies, indicating that the sufferers were incorporated into their own social groupings despite their conditions.

The traits selected to distinguish each burial from those adjacent graves are divided into those constituting the burial rite (body position, grave form, orientation and depth) and the grave goods (copper, polished stone ornaments, ground and polished stone tools, chipped stone, bone and antler, fired clay, pottery and animal bones). While the four burial traits are ubiquitous in each group, the number of grave goods categories in any one group varies from three to nine.

The westernmost Group (Group A: *Fig. 10*) comprises five graves – two female children at one end, a cenotaph in the middle and two adults, one female and the oldest male, at the far end. The only group trait is the standard burial orientation of West-East. The contrasts between each pair of graves can be measured against the total number of burial traits used in a given grave line (Table 12).

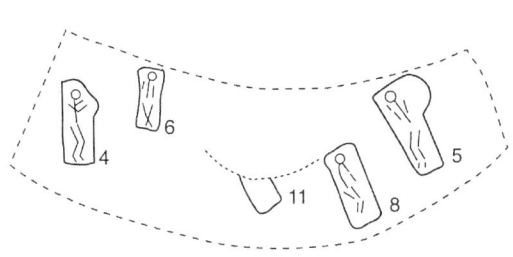

Fig. 10. Group A, Period I, Tiszapolgár-Basatanya.

Table 12. Grave good, costume and animal bone contrasts in Group A, Tiszapolgár-Basatanya.

Grave 4	Grave 6	Grave 11	Grave 8	Grave 5
copper bracelet	-	-	copper ring	-
bead girdle	bead girdle	bead girdle	bead girdle	-
flint blade	-	pebble	obsidian blade	flint blade
pebble				
pig bones	-	red deer teeth	-	pig mandibles
caprine skull				caprine bones

In Group A, the difference between each successive pair of graves falls between 60 and 70% of all traits (n = 10), emphasised mostly in grave goods. This suggests that a high degree of contrast was manifested to emphasise difference. Sexual differences are limited to the inclusion of bead girdles in all but male graves and the exclusion of copper from male graves. Sexual differences between the two adult graves are marked by grave form, copper, lithic raw material, and animal bones. No traits distinguish both children's graves from both adult graves but many traits separate one child's grave from the other: grave form and depth, copper, lithic types, pebbles, the number of pots and the animal bones. Those traits distinguishing graves where a person of the same age and sex is buried may be called 'individual' traits, recognising that social group and status may also be involved in this differentiation.

Next to Group A is the most artifact-rich grave line in the whole cemetery – Group B (*Fig. 11*). Characterised by a strong dominance of male graves (6/7 sexed graves, plus one cenotaph), with the oldest burial (the only female) at the end of the line, the group provides the greatest potential for inter-grave differentiation through the unusually wide range of grave good categories. As with Group A, the only group trait is standard burial orientation. The differences between successive pairs of graves cover a wide range of variability, between 40 and 80% of traits (n = 13) (Table 13).

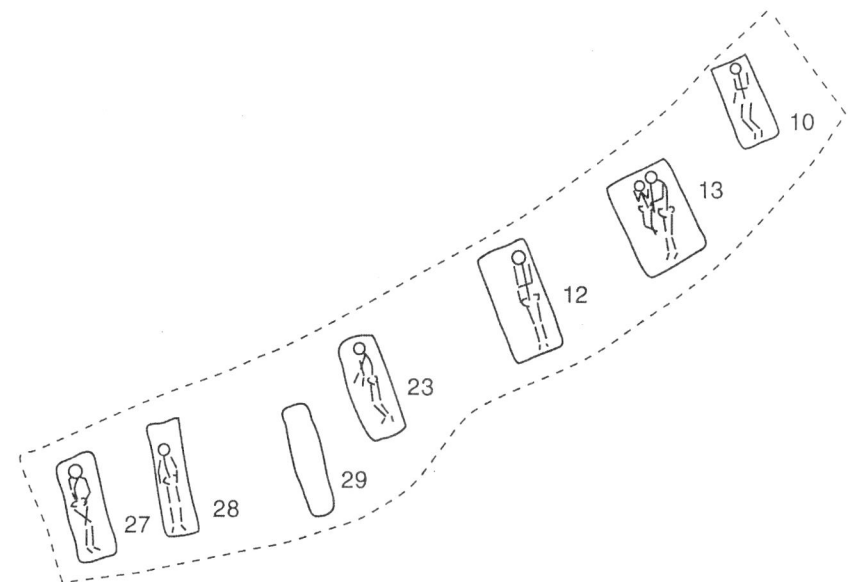

Fig. 11. Group B, Period I, Tiszapolgár-Basatanya.

Table 13. Grave good, costume and animal bone contrasts in Group B, Tiszapolgár-Basatanya.

Grave 27	Grave 28	Grave 29	Grave 23	Grave 12	Grave 13	Grave 10
-	copper beads	copper bracelet	bracelet + ring	copper bead	bracelet + bead	-
bead girdle	beads on chest	-	-	limestone disk	-	-
ground stone fragment	-	-	PS macehead	perf PS axe	-	-
				mini trap axe		
-	flint blades	-	flint blades	long flint blade	flint blades	quartz blades
			obsidian + flint	bag of tools	scraper	
			flakes			
-	antler fishpoint	cut antler	-	antler fishpoints	-	-
				bone awls		
caprine bones	-	caprine bone	caprine bones	caprine bones	caprine bones	-
-	pig mandible	boar's tusk	pig's tusk	pig's tusk	boar's tusk	pig mandible
			boar's mandible	pig mandible		
	aurochs scapula	aurochs scapula	-	aurochs scapula	aurochs scapula	-
		cattle metapodia				
-	hare bone	-	-	bird bone	red deer antler	

The contrasts are especially marked between female and male and between males and the sole cenotaph grave. The only female burial is contrasted with all the male burials by body position, grave form, bead girdle, grindstone fragment, mussel shells and the absence of flint blades and pig/boar bones. However, most contrasts define male burials of different ages. All grave forms were standard except for the irregular grave of the youngest, who also had the shallowest pit and lacked copper ornaments. Limestone beads were found with old and middle-aged burials, while a stone battle-axe and a miniature axe were placed with the oldest male and a mace-head with a young adult. With lithics, the youngest male was furnished only with quartzite tools whilst the oldest had a bag of tools and the only 'superblade' in the line. Equally, the widest range of bone and antler tools were placed with the oldest, together with lots of pots – a grave good set shared with the youngest burial. Many different combinations of domestic and wild pig bones denote different age ranges, especially tusks and mandibles. Aurochs scapulae were placed only with older males, while caprine bones were deposited with the oldest and the youngest. Other differences related to individual distinctions, such as those between the two 35-year-old males in body position, copper bracelet, limestone beads, antler fishpoint and different wild animal bones.

In Group C, a set of three female burials is followed by two older males (*Fig. 12*). The group traits are, positively, standard body orientation and, negatively, an absence of chipped stone grave goods. The only linear pattern is

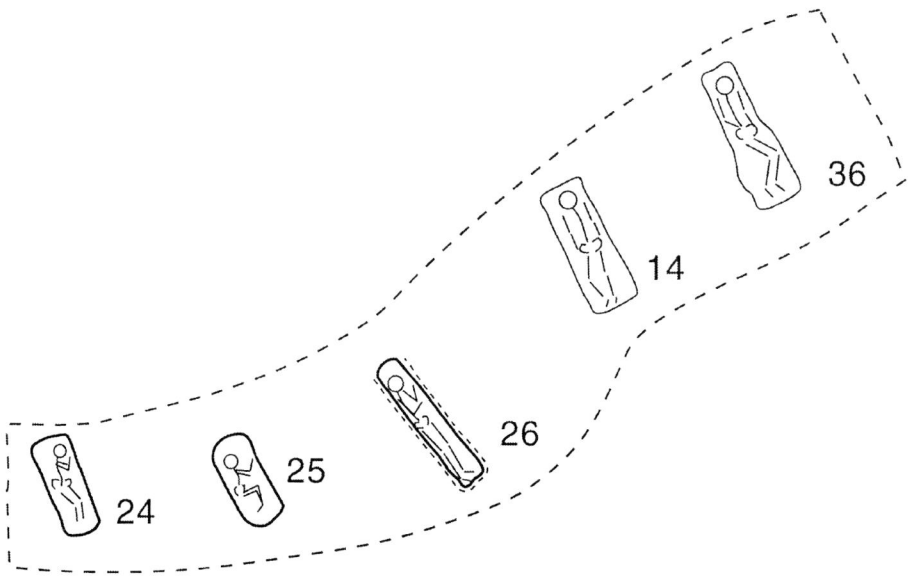

Fig. 12. Group C, Period I, Tiszapolgár-Basatanya.

an alternating increase and decrease in grave depth, correlated with the numbers of pots deposited. The scale of contrasts between successive pairs of graves varies between 40 and 70%, based on a wide range of burial traits (n = 12) (Table 14).

Table 14. Grave good, costume and animal bone contrasts in Group C, Tiszapolgár-Basatanya.

Grave 24	Grave 25	Grave 26	Grave 14	Grave 36
copper rings	-	-	-	-
bead, wire				
bead girdle	bead belt	-	-	-
-	-	-	mini trapezoidal axe	trapezoidal axe
-	-	pebble	-	-
-	-	-	loom weight	-
caprine bones	-	caprine bones	-	-
-	-	piglet bones	pig mandible	-
			boar's tusk	

The smallest divergences are for female children, while adult female:young female and old male: adult female attract the strongest contrasts. No traits separate all the female graves from all the male graves. There are, however, clear sexual differences between adults, in grave depth, polished stone axes and pebbles, which are counterbalanced by similarities in body position, grave depth, polished stone ornaments and the number of pots. Age differences between males occur in body position, grave form and depth, loom-weights and animal bones. The two female children's burials share body position, grave depth, stone ornaments and the number of pots, while contrasting in grave form, copper objects and animal bones.

The next line, Group D, is dominated by middle-aged persons, with the oldest burials at the ends of the line (*Fig. 13*). Two positive group traits occur: the rounded rectangular grave form (in 5 out of 6 burials) and the discovery of charcoal in the base of 4 out of 8 graves. A third possible trait is the deviation from the standard grave orientation, unusual in Period I and found in two graves here. In the successive pairs of burials, the greatest contrast (85%) lies between an adult female and a triple grave including a male of the same age (Table 15).

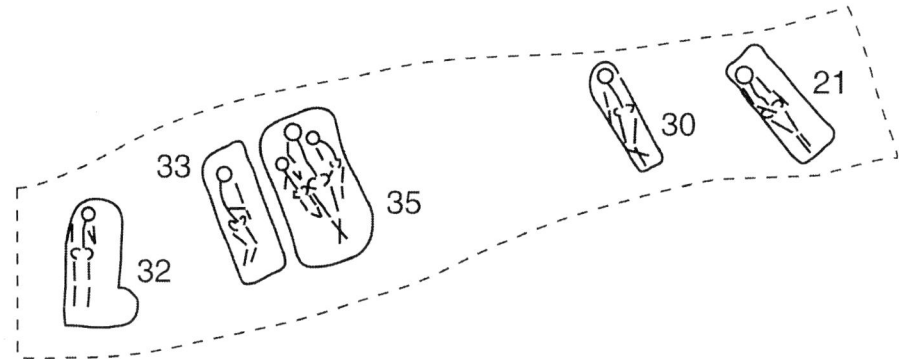

Fig. 13. Group D, Period I, Tiszapolgár-Basatanya.

Table 15. Grave good, costume and animal bone contrasts in Group D, Tiszapolgár-Basatanya.

Grave 32	Grave 33	Grave 35	Grave 30	Grave 31	Grave 21
-	copper ring	-	-	-	copper rings
-	bead belt	bead girdle	-	-	scattered beads
-	bead necklace				
-	-	GS fragment	-	-	-
-	pebble	flint blades	flint blades	flint flake	quartzite flakes
			flint flakes		
-	-	cattle metapodial			
		perforated boar's			
		tusk pendant			

Far less contrasts are drawn between adult males of the same age and adult males and infants (50%). The number of traits is the same as in Group C (n = 12). There are a few distinctions between all male and all female graves: the use of loose limestone beads in male graves, with bead belts in female graves, and the inclusion of fired clay lumps in female graves. But greater emphasis is placed on sexual differentiation between adult males and females, using body position, copper rings, lithic raw materials and types and animal bone species. Age variation within females graves is also accorded importance, based on body position and grave orientation, grindstone fragments and the number of pots (correlated as it is with grave depth). Finally, individual traits used to distinguish adult males include grave depth and form, boar's tusk pendants, cattle metapodia and aurochs scapulae.

Group E comprises a small all-male group with three burials of varying age and the oldest in the middle (*Fig. 14*). No clear group traits have been identified, since the material culture is drawn upon to emphasise age-related difference. A level of 70% contrast is used to differentiate between both burial pairs (n = 10) (Table 16).

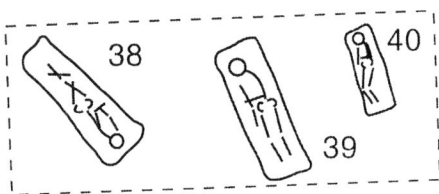

Fig. 14. Group E, Period I, Tiszapolgár-Basatanya.

Table 16. Grave good, costume and animal bone contrasts in Group E, Tiszapolgár-Basatanya.

Grave 38	Grave 39	Grave 40
-	copper bead	copper bracelet
-	-	scattered beads
flint blades	flint blades	flint blade + scraper
-	perforated shell disk	-
aurochs scapula	cattle metapodia	-
cattle tibia		
pig mandible	pig mandible	pig mandible
pig tooth	boar's tusk	

Two traits are shared in all three graves – flint blades and pig mandibles. The youngest is distinguished by a copper bracelet, polished stone beads and the least varied animal bone; the oldest by body position, copper beads, shell discs and the smallest number of pots. The middle-aged burial is the only grave in Period I oriented East-West, with a different grave form and depth from the other two and the inclusion of an aurochs scapula.

Group F is a larger group than E, also all-male but all middle-aged (*Fig. 15*). One probable group trait concerns the burial, with the humans, of a dog in each of the undisturbed graves; standard grave orientation is also practised. Age variation in adult males is demonstrated by the body position, grave form, lithic raw material, different numbers of pots and varying animal bone species. The individual traits separating the two 25-year-old males – both of whom met violent deaths – focus on the greater variety and quantity of grave goods in grave 52, with use made of body positions, grave depth, copper bracelets, superblades, antler and boar's tusk pendants, the number of pots, red deer antler, mussels and aurochs scapulae. There are more contrasting individual traits between these two graves (70%) than between older and adolescent males (50%) (n = 12)(Table 17).

Table 17. Grave good, costume and animal bone contrasts in Group F, Tiszapolgár-Basatanya.

Grave 20	Grave 51	Grave 52	Grave 53	Grave 42
-	-	copper bracelet	-	-
-	-	long flint blade	flint blade	flint blade
		flint blade		
		flint blade		
		obsidian arrowhead		
-	-	antler tool	-	-
		perf. antler hammer-axe		
		boar's tusk pendant		

The large Group G is a male-dominated line with the oldest burial at one end of the line (*Fig. 16*). The use of the standard body position and a standard orientation are the only identifiable group traits, occurring in nine out of eleven cases respectively. The relatively low level of contrast between graves, varying between 40 and 60%, may be a partial reflection of the similar age/sex categories of the deceased (Table 18).

95

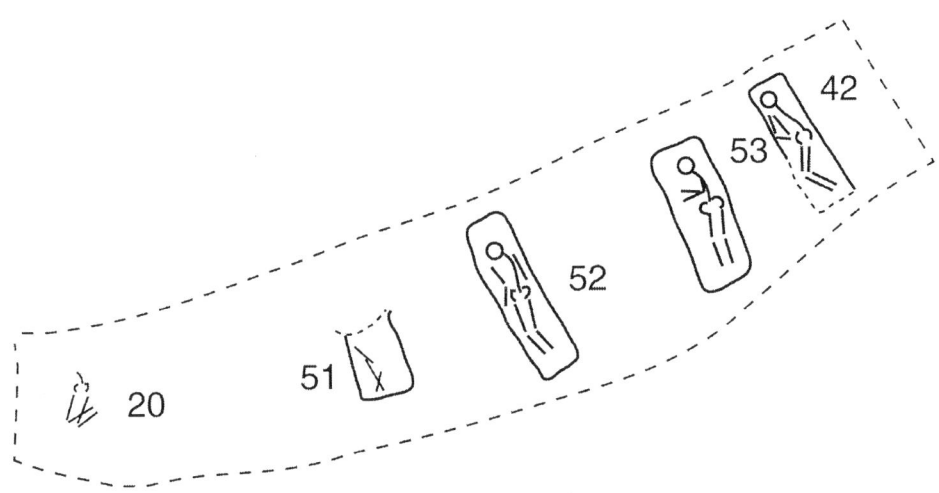

Fig. 15. Group F, Period I, Tiszapolgár-Basatanya.

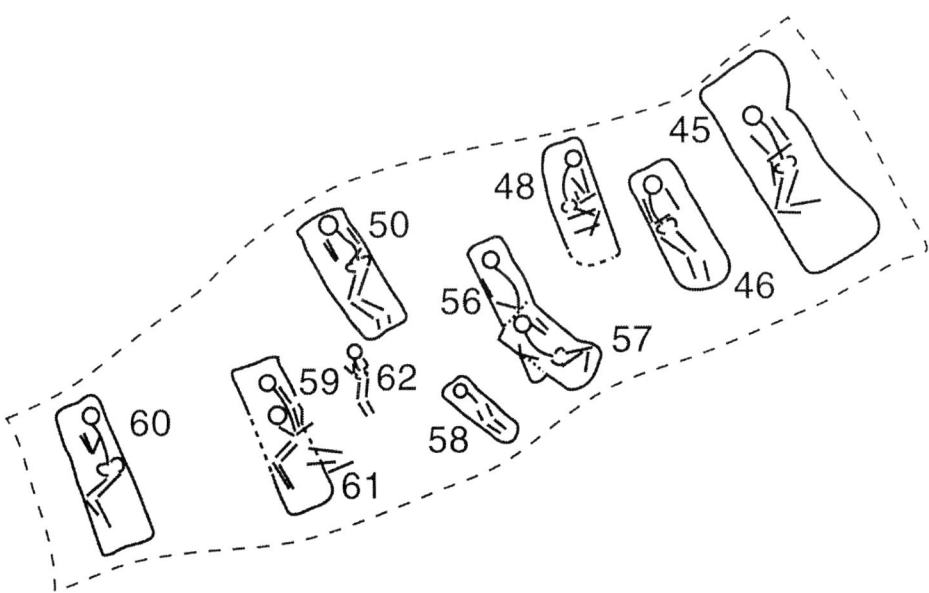

Fig. 16. Group G, Period I, Tiszapolgár-Basatanya (Burials 60, 61, 62, 50, 58, 56 and 46; later interpolations: Period I-II - 57, 48 and 45; Period II - 59).

Table 18. Grave good, costume and animal bone contrasts in Group G, Tiszapolgár-Basatanya.

Grave 60	Grave 61	Grave 62	Grave 50	Grave 58	Grave 56	Grave 46
-	-	-	copper bracelet	-	-	-
flint blades	flint blades	-	-	-	-	pebble
scrapers, flakes	flakes					
obsidian blade						
obsidian flakes	flake					
obsidian core						
antler hammer-axe	bone awl	-	-	-	-	-
cut antler						
boar mandible	pig mandible	-	pig mandible	-	pig mandible	-
pig mandible						
pig tusks, ribs						
aurochs scapula						
cattle bones	dog skeleton	-	-	-	carp vertebra	-
-	caprine bones	caprine bones	-	-	caprine bones	-

The greatest variation comes in the grave form, where there is a contrast between almost every pair of burials; a variant on this is the wave pattern of grave depth along most of the line, partly replicated in the presence/absence of pig mandibles. There is no single trait differentiating all three females and all the male burials. However, two out of the three female graves have unworked pebbles and all lack worked lithics, while the only graves with copper objects belong to adult males. While all three female graves contained a mussel shell, this is also found in a young male grave. Most of the grave line's variability is focussed on age differences amongst males. Children differ from older individuals in shallowness of grave pit, the small numbers of pots and an absence of lithics and pig mandibles. Young adults (18-20 years) differ through the deepest graves and presence of pig mandibles (although this is also found in the oldest male grave). The grave goods of the oldest adult stand out in diversity and quantity: far more lithics, the largest number of pots, the only boar's mandible and aurochs scapula, an antler hammer-axe and red ochre lumps. Individual variation within the same age-sex categories is limited among the children to grave form and presence/absence of sheep bones. Amongst young adults, there is far more individual variation – in grave form, the quantity of lithics and the presence/absence of bone awls, dog skeletons, sheep bones, snail and mussel shells, carp vertebrae and a horse tooth.

Group G is the only group where superposition of burials is attested, in two cases. While this evidence was discussed by Bognár-Kutzián in respect of the inner chronology of the cemetery (1963: 351-352), the two pairs could usefully be considered in relation to the contrasts drawn between the successive burials (Table 19). In each case, the grave of an older female partially covers the grave of a younger male.

Table 19. Contrasts between two pairs of superimposed burials in Group G, Period I, Tiszapolgár-Basatanya.

Graves 59 (later) and 61		Graves 57 (later) and 56	
SAME	DIFFERENT	SAME	DIFFERENT
Body position		Body position	
? Grave form			Grave form
Orientation			Orientation
	Copper ingot	No copper	
			Bead girdle
Obsidian flake	Obsidian core		Flint blades
	Bone awl		
Number of pots		Number of pots	
	Pig mandible		Pig mandible
	Dog skeleton		
	Caprine bones		Caprine bones
			Carp vertebrae

The age-sex differences between each pair of bodies makes identification of the factors causing variability quite difficult. It is interesting that, in one pair, the burial rites are much more contrastive, while difference based upon objects occurs at much the same frequency. The only object category used to express identity in both pairs is the number of pots. Otherwise, pig mandibles and caprine bones are used to express difference in both pairs, as are lithics. The ways in which material culture commonly in use is drawn upon to express identity and difference in these two pairs of superposed graves is no different from the use of material culture in successive non-overlapping graves.

The final Group in Period I is Group H (*Fig. 17*), also male-dominated, with three out of the four females at the North end of the line and the males to the South. There is also a general trend of older burials with time. The only group characteristic is burial in the standard orientation. The range of contrasts between pairs of graves is relatively low, from 33% to 65% (Table 20).

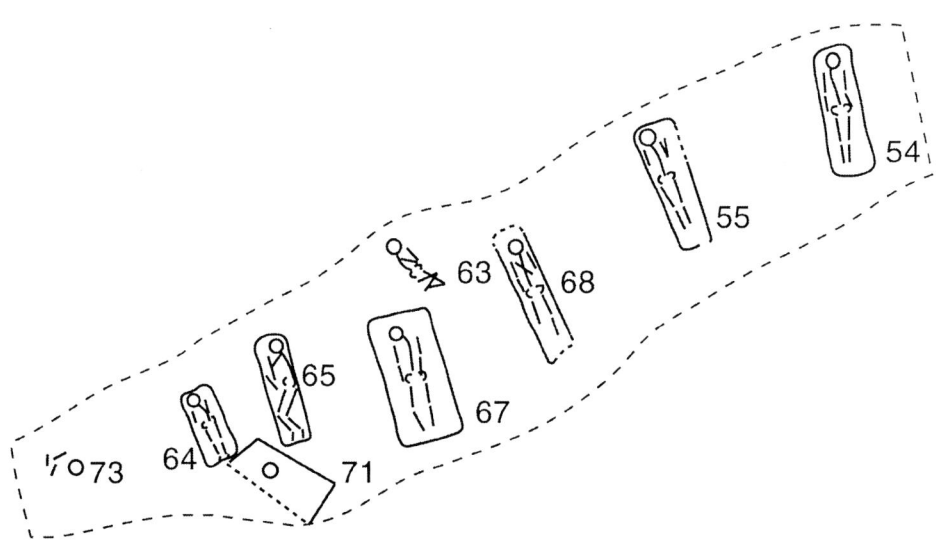

Fig. 17. Group H, Period I, Tiszapolgár-Basatanya (Burials 73, 64, 65, 67, 68, 55 and 54; later interpolations: Period I-II - 63; Period II - 71).

Table 20. Grave good, costume and animal bone contrasts in Group H, Tiszapolgár-Basatanya.

Grave 73	Grave 64	Grave 65	Grave 67	Grave 63	Grave 68	Grave 55	Grave 54	Grave 66
-	-	-	-	-	isolated bead	-	bead belt	-
-	-	-	flint blades	flint blades	-	-	-	pebble
			flint flakes					
			obsidian flake					
-	-	-	antler hammer-axe	-	-	-	-	-
			pol bone awl					
			pol antler point					
			worked human femur					
-	caprine bones	-	-	-	-	-	-	-
-	-	pig mandible	boar mandible	pig mandible	-	-	-	-
			pig bones	boar's tusk				
			dog skeleton					
			deer antler					
			aurochs scapula					
			cattle bones					

Grave depth varies in a wave pattern as far as the middle of the line, with more stability in the Northern part, partly correlating with the number of pots. There are no traits which distinguish all male from all female graves. The only traits separating male children and adolescents from mature adult males are the higher numbers of pots and the presence of lithics with the older males. Most differentiation comes between individuals of the same age/sex categories. Minor differences between male children/adolescents in grave form and depth and the use of sheep bones, while grave form, ornamental belts and unworked pebbles distinguish the older females. The greatest differences show the contrast between the two older males: body position, grave form and depth differ with obsidian and all bone/antler tools except one (a bone spoon placed with a mature female), a wider range of pig bones and the inclusion of a worked fragment of human bone all occurring with the older male.

4.4.2.1 Personhood and group identities

The Period I grave lines comprise a wealth of mortuary data which it is possible to partition into five different sources of variability. These are group traits, traits differentiating age-sex groupings, sex-based groupings and age-based groupings and traits defining individual differences. These categories are self-forming *sensu* Blake (see above, pp. 35-37).

In general, group traits are weakly developed in Period I, with only Group D exhibiting three traits. The commonest group trait – standard body orientation – is actually shared by five out of the eight groups and so its value for group differentiation is diminished. Group traits found uniquely (i.e. in one group only) are more valuable and include dog graves (group F), the rounded rectangular grave form (group D), extended burial on the back (group H), deviation from standard orientation (group D) and traces of burning in graves (group D). Group traits have the potential for making strong statements about difference; the lack of strong statements suggests that the links between members of different grave lines are as important as the separation into lines which, of itself, produces group identity.

There are only a few age/sex categories defined exclusively by one or more object categories. None of these applies to children and only one out of five exclusive categories applies to adult males. Adult females share limestone beads or belts made of beads as well as fired clay lumps in Group D, while bead girdles are restricted to adult females in Group A, in which adult males are defined by an absence of copper objects.

Rather greater differences exist for categorical differences in respect of sex, where age is not a relevant difference. In Group A, sexual difference is marked by differences in lithic raw materials (obsidian versus flint) and the presence for females only of pig mandibles and caprine bones. In Group B, bead girdles, grinder fragments and mussel shells were placed in female graves only, where there were no animal bones. In Group D, lithic raw material differences were noted alongside the presence of pig mandibles in male graves. However, in the five other Groups, no sex-related differences were noted.

Greater variability in burial rite and grave goods can be noted as a result of age-related differences. This occurred not only in Groups dominated by one sex (e.g., the all-male Group E and the male-dominated Group B) but also among females in Group D and males in Group C. In Group G, children were defined mostly by negative traits, while the oldest males were characterised by positive, mostly artifactual traits. Only in one Group (Group A) were there no recognisable age-related differences.

A last source of variability is the differences which emphasised the individuality of two or more burials of the same age/sex category. These differences were important in each Group except Group E, where there are no such pairs of individuals. Differences between children ranged from minor, in Groups C, G and H, to very significant in Group A. Adolescent males also showed major differences in Group G. Adult males exhibited more differences than adult females in Groups B, F and H, while adult females were more differentiated than males in Group D.

In summary, the complex pattern of variability in burial rite and grave goods deposition in Period I at Basatanya can be read as the material residue of human categorisation processes, in which five different sources of variability can be recognised. Surviving traits related to Group identities are less significant than those correlated to age-related differences and individual variations, which are, in turn, more highly differentiated than the age/sex categories and the sex categories.

4.4.3 Period II grave lines at Basatanya

The Middle Copper Age use of the Basatanya cemetery started with a series of grave lines in parallel to those of the Early Copper Age but offset to the South by almost ten metres. This decision may reflect knowledge of, and reverence for, the origins of the Early Copper Age cemetery on the Southern side of the hillock or may constitute a statement about distancing the group from the earlier grave line, or both at once. Whatever the reason, the end of the first Period II grave line touched the very edge of the Easternmost Period I line, without cutting any of the graves.

The Period II graves are grouped into a series of 12 lines, with a sizeable number of graves (n = 31 or about one-third) not included in these groupings. The groupings follows closely the Bognár-Kutzián ordering (1963: 219-221), with the following exceptions: Group I combines two small Bognár-Kutzián groupings; Group K omits the isolated graves 1 and 91, just as graves 3 and 82 are too far from Group L and graves 112 and 125 are off line; grave 111 is too far from Group M, graves 101 and 104 are too far from Group N, which includes 96 instead, and grave 106 is too distant from Group O; graves 113 and 119 form a pair rather than a Group, as do pairs of graves 126 + 140, 134 + 139 and 128 + 135; the trio of graves 124 + 144 + 146 are too distant from each other and, finally, the quartet of graves 133 + 132 + 136 + 143 are not close enough together to form a coherent Group.

This leaves the following distribution of age-sex categories for the 12 lines (Table 21):

Table 21. Age/sex composition of grave lines and other graves, Period II, Tiszapolgár-Basatanya.

Group	Adult Female	Adult Male	Female Child	Male Child	Incomplete Information	Cenotaph	Total (Path.)
I	2	4	-	-	-	-	6 (3)
J	4	4	-	1	-	-	9 (2)
K	2	1	-	-	-	-	3 (2)
L	3	3	1	-	-	-	7 (3)
M	3	1	-	-	-	-	4 (3)
N	2	2	-	1	-	-	5 (2)
O	3	-	-	2	-	-	5 (-)
P	1	2	-	1	-	-	4 (3)
Q	5	2	-	1	-	-	8 (3)
R	1	2	-	-	-	-	3 (3)
S	-	5	-	-	-	-	5 (3)
T	3	-	-	1	-	-	4 (2)
Sub-T	29	26	1	7	-	-	63 (29)
Other	11	14	-	1	5	-	31 (1)
TOTAL	40	40	1	8	5	-	94 (30)

Key: (path) = number of skeletons with a pathological condition.

The mean number of burials in lines is similar within the range of error (6.0 in Period I, 5.2 in Period II). There are, however, three main differences between the lines of the two Periods: the greatly reduced number of Period lines with adult males, adult females and children (down to 50% of lines); a lower fraction of Period II graves placed in lines (from 78% to 67%); and a smaller percentage of Period II burials with symptoms of pathology (from 40% to 32%). Let us now turn to an analysis of each grave lines to see how local differences are developed in a period with a weaker attachment to the traditional performative rules so vigorously maintained in Period I.

In Group I (*Fig. 18*), there is a dominance of adult males over females, with children excluded; while three of the males are mature (over 40 years), neither of the females is over 30 years. The scale of contrasts between successive pairs of graves varies between 20 and 70%, based on a wide range of burial traits (n = 11) (Table 22).

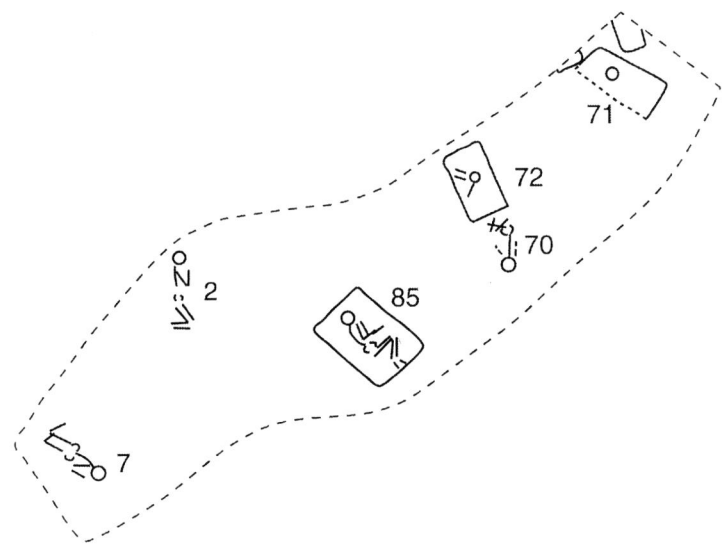

Fig. 18. Group I, Period II, Tiszapolgár-Basatanya.

Table 22. Grave good, costume and animal bone contrasts in Group I, Tiszapolgár-Basatanya.

Grave 7	Grave 2	Gravev 85	Grave 70	Grave 72	Grave 71
-	-	-	-	-	copper awl
-	-	isolated beads	-	-	-
-	-	flint blade	-	flint blade	flint blades
					obsidian arrowhead
-	-	bone awl	-	-	-
-	-	pig mandible	-	-	cattle bones
		pig bones			
		caprine bones			

The smallest difference occurred between an adult male and female. The only possible group trait is left-sidedness, which occurs in five out of the six burials, despite the number of males. There is a linear decrease in grave depth to the North, with the exception of the last grave (71). Age-based contrasts cannot be investigated because of poor preservation in the young male grave. There are no sex-based differences found in all cases but the non-standard orientation in three out of four males contrasts with the normal orientation in both female graves. Individual differences are more striking. Although similar in burial rite, the females exhibit strongly contrastive grave goods (ornaments, tools, pottery and animal bones). The differences between the mature males are reinforced in all aspects of burial rites as well as many grave goods.

In the large Group J (*Fig. 19*), there is a sexed balance in adult graves plus a single child burial. The level of contrasts between successive graves varies between 40% and 70%, on the basis of 10 traits; the smallest difference occurred between an adult male and a female (Table 23).

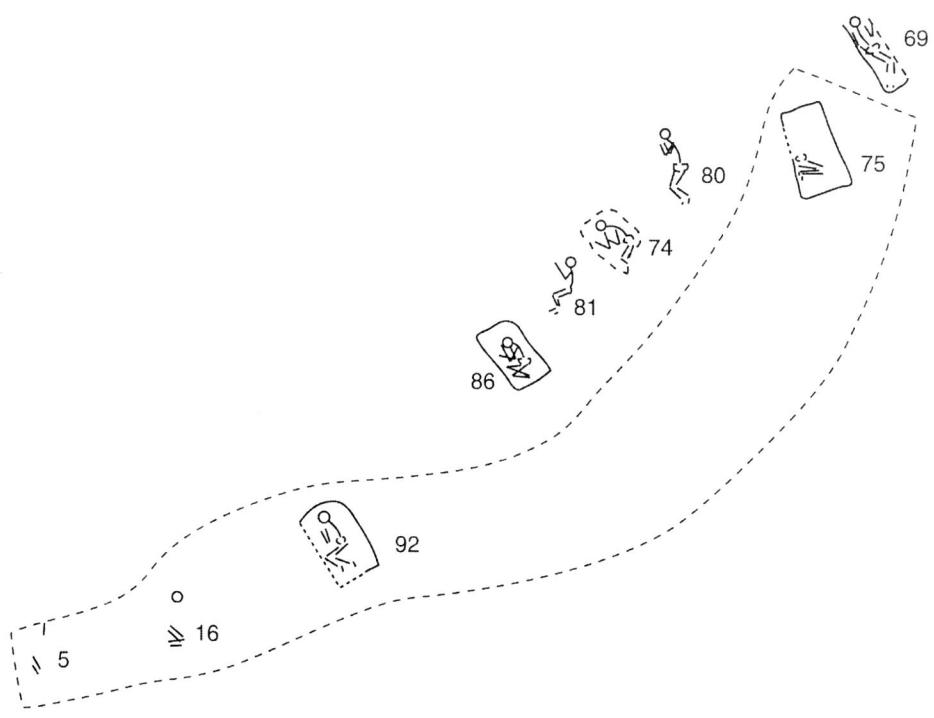

Fig. 19. Group J, Period II, Tiszapolgár-Basatanya.

Table 23. Grave good, costume and animal bone contrasts in Group J, Tiszapolgár-Basatanya.

Grave 5	Grave 16	Grave 92	Grave 86	Grave 81	Grave 74	Grave 80	Grave 75	Grave 69
-	-	copper awl	-	-	pin	-	-	-
-	-	-	-	-	-	-	-	isolated beads
flint blade	-	flint blade	-	-	flint blade	-	-	-
-	-	obsidian core	-	-	obsidian lump	-	-	-
pig mandibles	-	pig bones	-	-	pig bones	pig mandible	pig bones	caprine bones
caprine bones	-		-	-		pig bones	antler	antler
						cattle bones		

The only possible group trait is the standard West-East orientation, found in seven out of nine cases. Instead, grave forms are used to emphasise contrasts and there is a linear trend in grave depth, deeper at the ends and shallower in the middle of the line. There are neither age- nor sex-related differences affecting all cases, but two sex-based trends occur – the association of chipped stone with most adult males and the deposition of pig bones with all males but also with the oldest female. Age-based contrasts among males are found in the variations in grave form and depth and the presence/absence of copper, lithics, the number of vessels and the kind of pig bones. Individual differences are more marked in adult females, through body position, grave form and depth and the deposition of limestone beads, animal bones and varying numbers of vessels. The only individual difference for younger males is in the presence/absence of lithics.

Group K is a small group of three burials – two young adult females and a mature male (*Fig. 20*). Contrasts between these burials remains at a high level of 80% for both pairs (Table 24).

Table 24. Grave good, costume and animal bone contrasts, Group K, Tiszapolgár-Basatanya.

Grave 93	Grave 87	Grave 83
-	copper bracelets	-
limestone bead girdle	calcite bead girdle	-
-	flint blade	flint blades
	pebble	obsidian core
		obsidian arrowhead
-	fishbone tattooer	-
	bone awl	
-	red ochre	-
	calcareous lump	

The group traits comprise the standard orientation, a linear increase in grave depth to the North and the negative trait of the absence of animal bone deposition. In small groups of this kind, contrasts can be explained in one of two ways: age-sex differences (between females and male) and individual differences (between the two females). Both are emphasised strongly, the former by grave depth and the deposition of bead girdles, obsidian and a larger number of vessels. Individual differences are stated through variations in body position, grave form and depth and through the presence / absence of copper ornaments, lithics, bone awl, fishbone tattooer and colouring material, as well as the use of different materials for bead girdles.

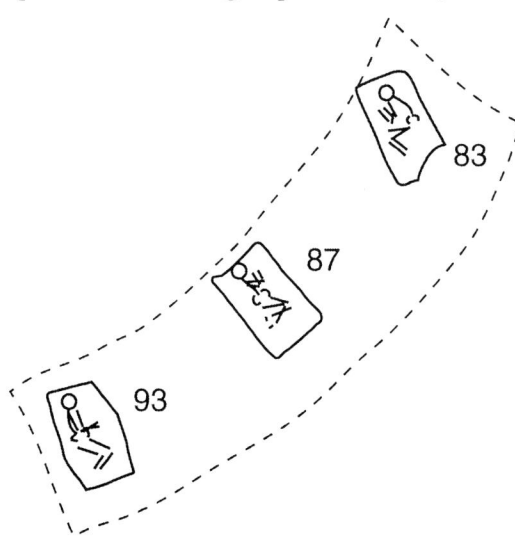

Fig. 20. Group K, Period II, Tiszapolgár-Basatanya.

Group L is dominated by adults of both sexes in equal numbers (*Fig. 21*). Contrasts between successive burials vary from 20% to 90% on a trait list of nine, with high values between adult males and females, low values between females of the same age (Table 25).

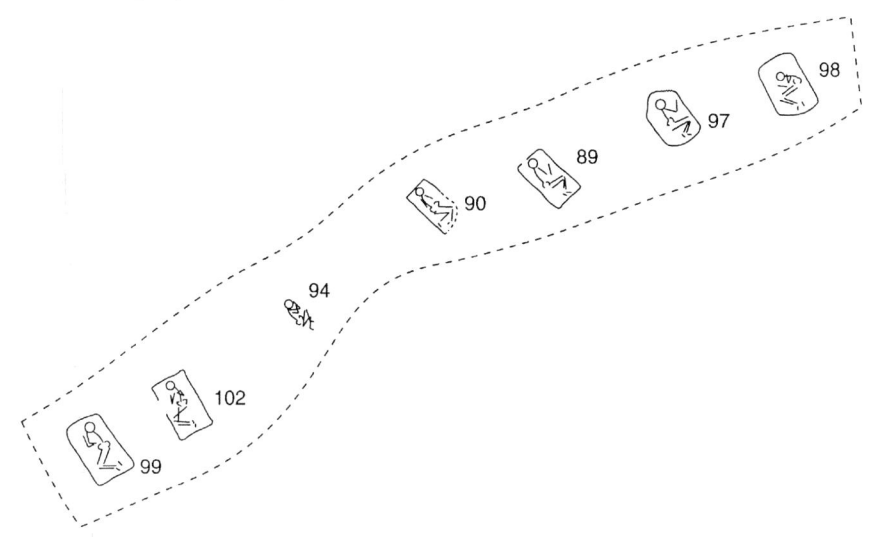

Fig. 21. Group L, Period II, Tiszapolgár-Basatanya.

Table 25. Grave good, costume and animal bone contrasts in Group L, Tiszapolgár-Basatanya.

Grave 99	Grave 102	Grave 94	Grave 90	Grave 89	Grave 97	Grave 98
copper awl	-	-	-	-	-	copper awl
-	-	-	-	-	-	ground stone + rubber
flint blade	flint blades	-	flint blade	-	-	long flint blade
	flint scraper					pebble
pig bones	pig bones	-	-	metapodia	-	boar's tusk

Group traits are defined by standard body position and orientation and the negative trait of the absence of polished stone ornaments. Grave depth follows a wave pattern, totally unrelated to the number of vessels. Compared to Groups I and J, this is a young group, with the oldest males, at 35 years, placed at either end of the line. The standard body position distinguishes all males from all females, as does the low number of vessels with females and the pig bones with male burials. Partial differentiation is found with copper ornaments (two out of three males, no females) and chipped stone (all males and the oldest female). The only age difference among the males is the presence of copper with the oldest. Female age-based contrasts are more common, including grave form and the number of vessels, as well as the absence / presence of flint blades and animal bones.

The small Group M (*Fig. 22*) comprises three adult females and one male of very similar ages (35-40 years). As well as the standard body position and orientation, the group is defined by the irregular rectangular grave form, found in each grave where the evidence is preserved. The contrasts in successive burials vary from 10% (between two females of similar ages) to 80% (between a male and a female of the same age), on a trait list of 10 (Table 26).

Table 26. Grave good, costume and animal bone contrasts in Group M, Tiszapolgár-Basatanya.

Grave 116	Grave 105	Grave 100	Grave 96
-	copper dagger	-	-
	copper pin		
-	ground stone fragment	-	-
-	long flint blade	-	-
	flint blade		
	flint core		
	obsidian core		
-	boar's tusk	-	-
	caprine bones		

In view of the shared ages of the buried, the main differences are age-sex-based and individual. The former comprise body position and the male association with copper, ground and polished stone, lithics and animal bones. Individual differences between females are not so striking, being restricted to grave depth and the number of vessels deposited.

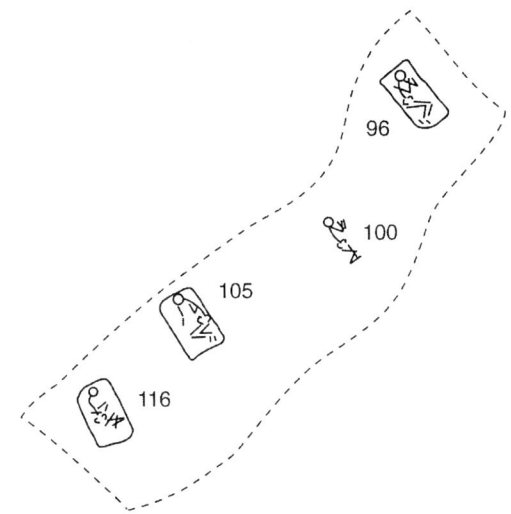

Fig. 22. Group L, Period II, Tiszapolgár-Basatanya.

Group N is composed of four graves and five bodies – a double burial with a female and child, two males and another female (*Fig. 23*). Group traits are marked by standard body position and orientation, as well as the absence of a single standard grave form. The contrasts between successive burials vary from 30% (between the double burial and the oldest male) and 90% (between adult males and females) (Table 27).

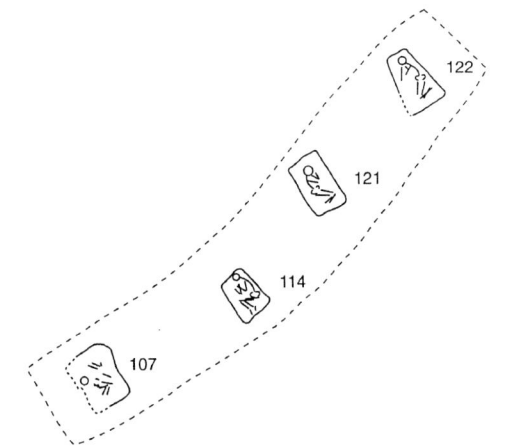

Fig. 23. Group N, Period II, Tiszapolgár-Basatanya.

Table 27. Grave good, costume and animal bone contrasts in Group N, Tiszapolgár-Basatanya.

Grave 107	Grave 114	Grave 121	Grave 122
-	-	bead girdle	-
flint blade	-	flint blade	flint blade
		obsidian blade	
		pebble	
-	-	bone awl	-
-	-	spindle whorl	-
-	-	pig mandible	-
		pig bones	
-	-	red ochre	-

The alternating burial of males and females is emphasised in the wave pattern of grave depth (female graves being deeper) and in the variations in the number of pots (more with females). Age-based differences are minimal with males (grave form and depth, flint blades) and very important between females. Apart form the grave form, the presence / absence of bead girdles, obsidian, unworked pebbles, bone awls, spindle-whorls, animal bones and colouring relates to contrasts with a single well-furnished female grave. Pots and flint blades are the only objects shared by all three females.

Three adult females and two males children comprise the medium-sized Group O (*Fig. 24*). Group traits include the standard body position and a tendency towards non-standard orientation (in three out of the five cases). The contrasts between successive burials vary from 50% to 75%, using 12 traits, the greatest contrast occurring between two females of the same age (Table 28).

The alternating deposition of male child and adult female is reinforced only by variations in body position, grave depth (deeper for male children) and the number of vessels (larger for females). Age-based differences between adult females concern grave orientation and depth as well as the presence / absence of unworked pebbles and colouring material. Individual differences between the younger females are widespread: grave form and the presence / absence of human bone deposits, bead girdles, ground stone rubbers, carp tattooers, carp vertebrae and pig bones.

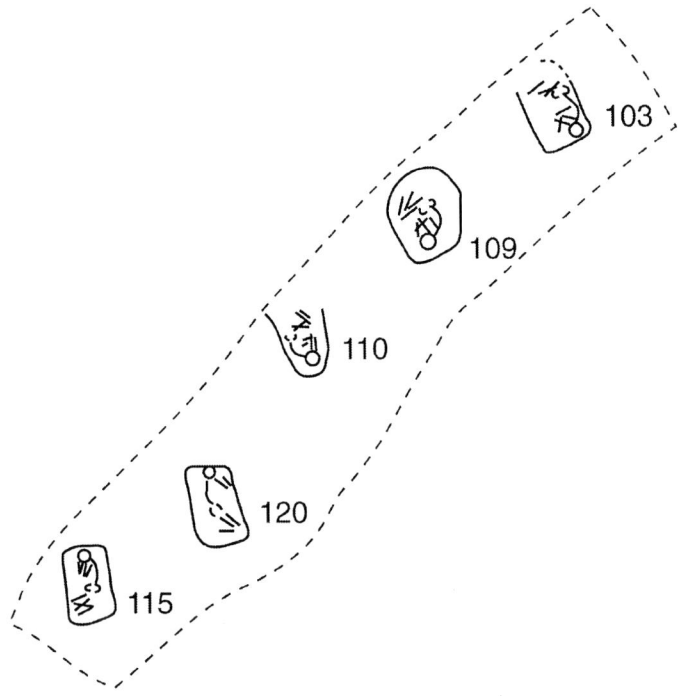

Fig. 24. Group O, Period II, Tiszapolgár-Basatanya.

Table 28. Grave good, costume and animal bone contrasts in Group O, Tiszapolgár-Basatanya.

Grave 115	Grave 120	Grave 110	Grave 109	Grave 103
-	-	-	-	human femur frag.
-	-	-	bead girdle	-
-	-	-	stone polisher	-
-	pebble	-	-	-
-	-	-	carp tattooer	
-	-	-	pig mandible	rib
			pig bones	
			carp vertebra	

Group P is a small group with adult males, an adult female and a child (*Fig. 25*). Three of the bodies have pathological conditions, including one violent death, and the removal of another's feet. There are three group traits: standard body position and orientation and the provision of flint blades in all graves, even the female burial. Contrasts between successive burials vary from 30% (between adult male and male child) and 70% (between adult female and male), on the basis of 10 traits (Table 29).

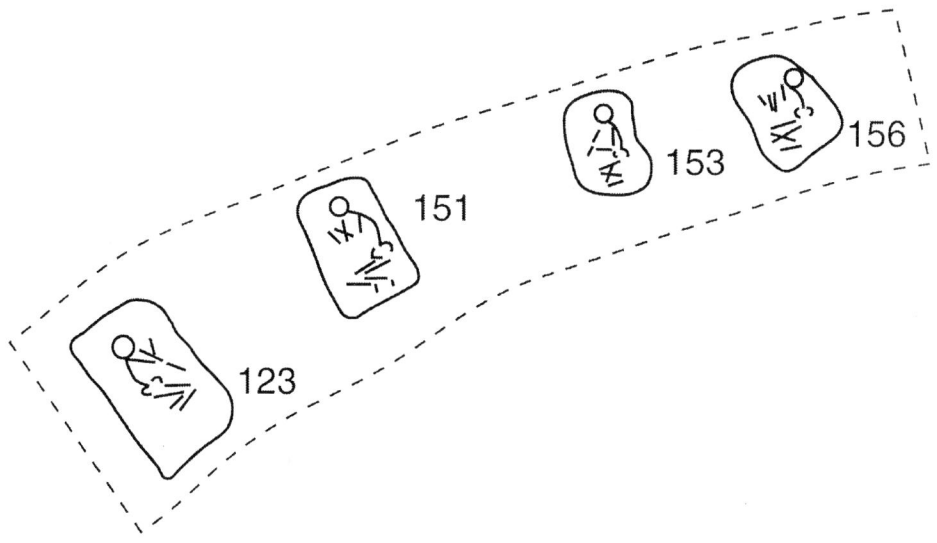

Fig. 25. Group P, Period II, Tiszapolgár-Basatanya.

Table 29. Grave good, costume and animal bone contrasts in Group P, Tiszapolgár-Basatanya.

Grave 123	Grave 151	Grave 153	Grave 156
bead girdle	-	-	-
flint blade	flint blade	long flint blade	flint blade
pebble	obsidian core	flint blade	
bone awl	-	-	-
caprine bones	-	-	pig bones

Age-sex differentiation is based upon associations with the adult female, including the bead girdle, unworked pebble, bone awl and caprine bones, as well as a larger number of pots. Age-based differences among the males were marked in grave form and depth, lithics types and the presence / absence of pig bones.

Group Q is a large group of burials (*Fig. 26*), dominated by adults (with only one child), of which the majority were females. There was a larger fraction of mature adults than usual, with half of the burials aged between 40 and 60 years. The only group traits were the standard body position and orientation. The contrasts between successive burials varied between 10 % and 70%, based upon 10 traits: the smallest variation between a young female and a male child, the greatest between a mature male and the same young female (Table 30).

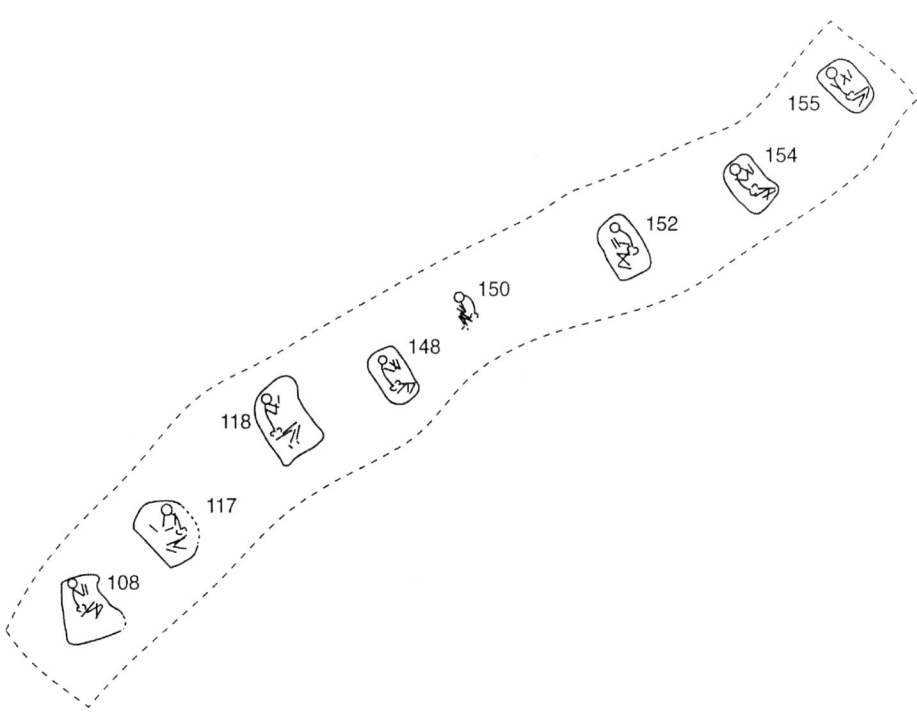

Fig. 26. Group Q, Period II, Tiszapolgár-Basatanya.

Table 30. Grave good, costume and animal bone contrasts in Group Q, Tiszapolgár-Basatanya.

Grave 108	Grave 117	Grave 118	Grave 148	Grave 150	Grave 152	Grave 154	Grave 155
-	human femur	-	-	-	-	-	-
-	copper pin	-	-	-	-	-	-
-	flint flake	pebble	-	-	flint blade	-	-
	quartz flake						
caprine bones	caprine bones						
-	-	-	-	-	-	-	fired clay nail

Contrasts referring to all the older males versus females were restricted to lithics and the number of vessels; such differences did not exist between all younger males and females. Likewise, there were no overall sex-based differences relating to all members. Instead, age-based variations were found in both females (grave depth) and males (presence / absence of human bone deposition, copper pin, lithics and caprine bones). Individual differences among females can also be separated into young females (grave form, the number of vessels, presence / absence of unworked pebbles) and older females (grave form and the presence / absence of caprine bones and the fired clay nail).

The small Group R comprises three adults (*Fig. 27*) – one female buried between two males. No group traits can be distinguished in this group, where the contrasts between pairs of graves is high (70-100%) on the basis of seven traits (Table 31).

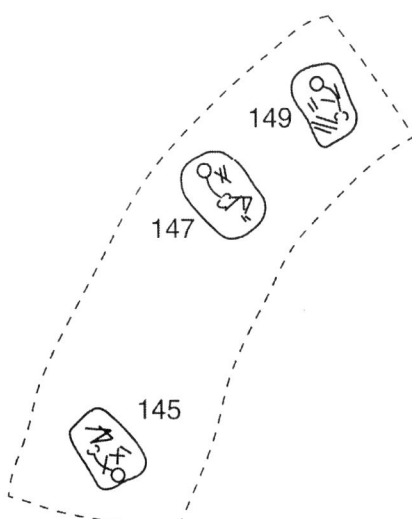

Fig. 27. Group R, Period II, Tiszapolgár-Basatanya.

Table 31. Grave good, costume and animal bone contrasts in Group R, Tiszapolgár-Basatanya.

Grave 145	Grave 147	Grave 149
flint blade	-	flint blade
caprine bones	-	-

Sex-based differences are expressed in grave form and depth, the number of vessels and the presence / absence of flint blades. Age-based contrasts between the males amount to body position, orientation and the presence / absence of animal bones.

The medium-sized Group S (*Fig. 28*) is composed of nothing but adult male burials, all of whom are over 35 years. The only group trait is a non-standard, East – West orientation, found in three out of the four cases where the evidence survives. The contrasts between successive graves range from 25% to 80%, on the basis of 12 traits: the lowest between two adult males, the highest between adult males of different ages (Table 32).

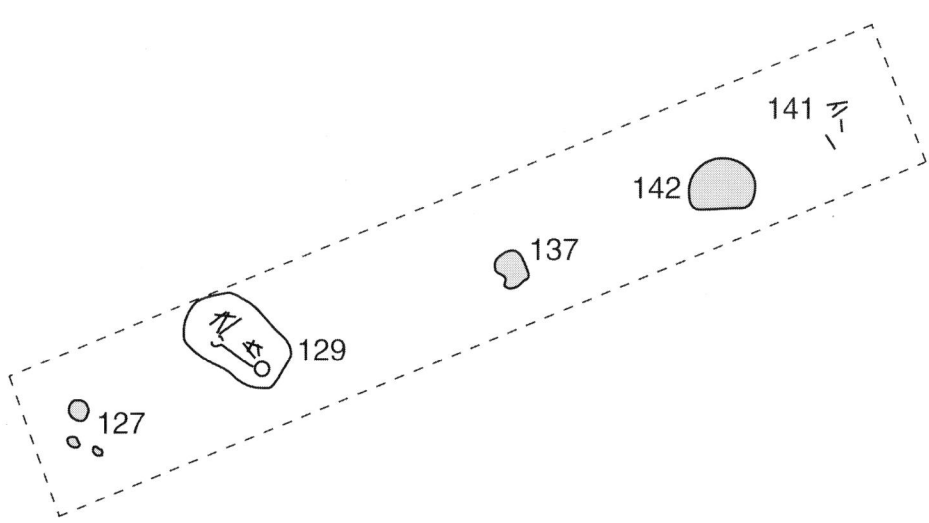

Fig. 28. Group S, Period II, Tiszapolgár-Basatanya.

Table 32. Grave good, costume and animal bone contrasts in Group S, Tiszapolgár-Basatanya.

Grave 127	Grave 129	Grave 137	Grave 142	Grave 141
-		-	-	gold bead
-	copper pin	-	-	-
-	long flint blade	-		obsidian blade, flint blade
-	flint nodule			obsidian arrowhead
-	perforated stone battle axe	-	-	-
	caprine bones cattle scapula			-
-	red ochre	-	-	-

Age-based contrasts are strong in this group. The youngest male was the only burial placed on the left side with standard orientation, accompanied by the largest number of vessels and by cattle bones. The oldest male, buried in the deepest grave, was associated with copper, a superblade, polished stone battle-axe, caprine bones and red ochre. The only gold object in the whole cemetery – a small bead – was placed with an adult male of uncertain age.

Four burials from three graves comprise the Easternmost Group T (*Fig. 29*) – a double grave with an adult female and a male infant and two other adult females. The contrasts between the two pairs of graves reach the 70% level in both cases (Table 33).

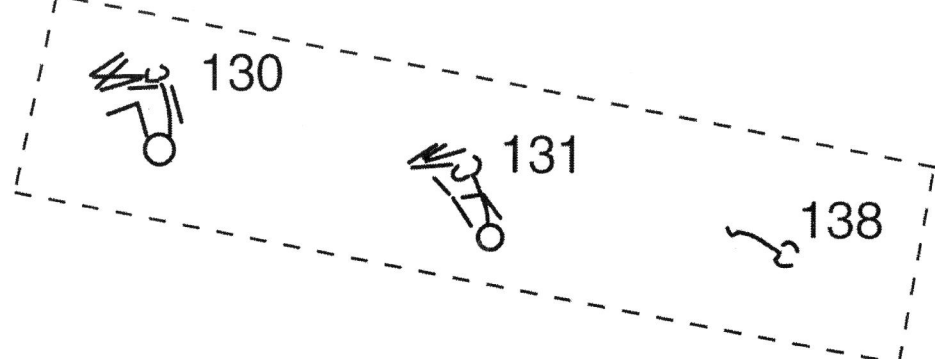

Fig. 29. Group T, Period II, Tiszapolgár-Basatanya.

Table 33. Grave good, costume and animal bone contrasts in Group T, Tiszapolgár-Basatanya.

Grave 130	Grave 131	Grave 138
bead girdle	scattered beads	-
-	red ochre	-

Group traits stand in contrast to one another: standard body position and non-standard orientation. The only trait differentiating the male infant from the adult females is the presence of red ochre. There are no age-based differences between the females. Young females are differentiated by grave depth and orientation, the number of vessels and the presence / absence of a bead girdle.

4.4.3.1 Personhood and group identities

As in the Period I grave lines, Period II mortuary deposits exhibit a huge amount of variability. The same form of categorical analysis is used as with Period I graves, with a division of traits relating to five classes of information: group traits, age-sex-based traits, sex-based traits, age-based traits and individual traits.

There is only one grave line in Period II which is not defined by a group trait – the small Group R. Most of the remaining Groups are characterised by two or three traits, with only three Groups having a single trait. However, the same two traits are common to many Groups – the standard body position (7 Groups) and the standard grave orientation (also 7 Groups). The multiple occurrence of these traits obviously limit their utility for the definition of group uniqueness. However, only two of the Groups utilising these standard practices are characterised by these traits alone – Groups N and Q. Three of the remaining group traits are defined in opposition to these standard practices – left-sidedness (Group I), right-sidedness (Group J) and non-standard orientation (Groups O, S and T). In all cases of the last-named, the preferred mode is East-West, reducing the value of this trait for group definition. The other burial traits concern the choice of irregular rectangular graves by Group M and the linear trend to increased grave depth of Group K. A single positive group trait based upon objects is found – the ubiquity of flint blades in Group P, while two negative group traits may be proposed – the absence of animal bone deposition in Group K and the absence of polished stone ornaments in Group L. This pattern suggests that the definition of group traits in Period II was more important than in Period I and was achieved more successfully by drawing upon a wider range of burial practices and material emblems.

The categorical differentiation of age-sex groupings was rather more important in Period II than it was in Period I. Omitting the all-male Group S by definition, eight out of the remaining 11 Groups included age-sex-related object categories, while one Group (N) also utilised grave depth variation in this way. Only one object category was uniquely related to the age-sex category of children – the red ochre colouring in Group T. The commonest object category was the number of vessels, used in five out of eight Groups, usually to differentiate adult males from females. Adult males were uniquely associated with obsidian cores (Group K), pig bones (Groups L and M) and copper and ground stone (Group P). Adult females were, conversely, solely associated with bead girdles (Groups K and P) and unworked pebbles, bone awls and caprine bones (Group P). A wide range of material culture was drawn upon to validate such age-sex differentation.

In comparison with Period I, the association of objects and burial rites with specific sex-based categories of people, exclusive of age, was very restricted in Period II. With the exception of Groups S and T, where single-sex groupings are found, only three out of the 10 remaining groups exhibited any sex-based categorisation processes. The most complex Group in this respect is Group L, where three material contrasts are drawn: copper awls and flint blades with adult males and differences in grave form between adult males and females. In Group Q, larger numbers of vessels were deposited with adult females, while lithics were found only with adult males. Sex-based differentiation in Group R was based upon contrasting grave forms and depth. It is interesting that the apparently significant social discrimination through sex-based categorisation processes observed in the total sample of Period II graves by Bognár-Kutzián, Sofaer Derevenski and Chapman does not appear to be equally important over the whole of the Period II cemetery.

The widespread use of age-related differentiation stands in contrast to the paucity of sex-related categorisation. Although absent in four out of the 12 grave lines, age-based differentiation is significant in all of the others. It is interesting to note that there is no Group in which age-related differentiation occurs with females graves alone. Rather, male graves are contrasted by age-based categorisation processes in four cases (Groups J, P, R and T), while both female and male graves are contrasted by their age in four Groups (L, N, O and Q). We shall start with these latter Groups.

In Group L, the only object used for male age-differentiation is copper, with most of the traits being restricted to the female graves. In Group N, the only burial practice used to make age-related distinctions for both males and females is the grave form. In all categorisations using objects, different artifacts are selected for male and for female age-differentiation, with far more objects used for female grave goods. The same is true for Group O, where grave orientation is the only practice used for both male and female age-differentiation. In Group Q, no objects nor practices are common to both male and female age-differentiation and the wider range of objects is related to male graves. In the Groups with age-differentiation restricted to male graves, grave form and depth, together with copper, lithics and animal bones are most commonly used as cultural resources for age-based identities.

The final source of mortuary variability concerns individuals matched in age and sex whose burial rites and grave goods serve to differentiate the deceased as distinct persons. There is no evidence for individual variability in five out of the 12 Groups. Nonetheless, in most of the other Groups, widespread use is made of material culture to emphasise these contrasts. In two Groups (I and J), both male and female individuality is exhibited; in all the remaining Groups, female individuality is emphasised, whether for younger females (Groups O and T) or younger and older females together (Group Q). The range of material usage ranges from restricted (as in Group M) to very intensive (Groups I, K and O). The concentration on female individuality stands in marked contrast with the male-dominated age-differentiation discussed above and therefore can be seen to make a major contribution to social identity in the mortuary domain.

In summary, the remarkable mortuary variability recorded for the second period at the Basatanya cemetery can be fully dissected using categorical analysis. The same five forms of categorical differentiation can be identified as were found in Period I, but the relative importance of these traits in Period II is rather different. As a qualitative judgement, group identities were stronger in Period II than in Period I, as were age-sex-related and individually-based differentiation. Sex-based differentiation was less significant in the second Period than in the first, while the relative importance of age-related variations seems to have remained at the same level. These results make it possible to attempt the characterisation of the relative importance of all of the five forms of variability for each individual Group of graves.

4.4.4 Discussion

The burial of such a large number of people, by the standards of Copper Age times, at Basatanya is, first and foremost, the sign of a successful integration of households into the wider community. If failed lineages are, by definition, archaeologically invisible, the Basatanya cemetery is conspicuously successful, whether its use-life was 200 or 900 years. It must be admitted that this investigation has not shed light on the length of the cemetery's use-life.

This crucial information is vital to any nuanced interpretation of the grave Groups and their usage and holds this analysis back from a realisation of its full potential. The uncertainty over the mean interval between successive burials, with estimates ranging from one year to five, hinders an understanding of social memory and inhibits our understanding of the significance of community elders. If burials were a regular, annual event, most middle-aged individuals could recall the practices relating to over 20 funerals; if burials were spaced at five-year intervals, only the elders could build up a consistent picture of mortuary practices against which to measure the next funeral. A longer interval between funerals would suggest the likelihood of greater variability between successive burials, all other things being equal.

One quantitative measure of variations between successive burials is the range of contrasts between such graves by Group (*Fig. 30*). The range of Group variations is not normally affected by either the size of the Group (with the exception of the small Groups E, K, R and T) or the number of traits available for differentiation (except the small Groups R and T). In these small Groups, there is a heightened emphasis on difference, with one pair of graves in Group R differing on every single trait! All other Groups draw upon a broadly similar range of burial rites throughout the life of the cemetery and a range of similar objects during each Period and sometimes in both Periods.

There is a clear difference between the uses of material differentiation in the two Periods at Basatanya. In Period I, there is only one grave out of 60 in which 40% of all traits were similar to the preceding burial. By the same token, there were few graves whose traits differed from the preceding burial by more than 70%. This 'middle way' characterises the overall (*post hoc*) pattern of the establishment of difference in Period I. The extremes of close similarity and widespread difference are more often encountered in Period II. Here, a significant number of graves differ from their neighbours in only one or two traits (10-20% difference), while there are graves exhibiting 80% difference in half of the Groups. However, the commonest variation of around 40-60% for the majority of graves is shared by both Periods. It is clear that differentiation of most burials from the preceding event is a long-term cultural practice, developed at the earliest stage of the cemetery's use-life and carrying on throughout.

At the time of the earliest burials at Basatanya, there was, by definition, no local precedent for the conduct of funerals – rather, a cultural tradition developed in the course of an unknown length of time by earlier settlers South of Polgár, perhaps including the inhabitants of the Bosnyak domb tell in the Late Neolithic period. The earliest burial on the site – also dating to the Late Neolithic – provided a spatial reference point as well as a cultural standard

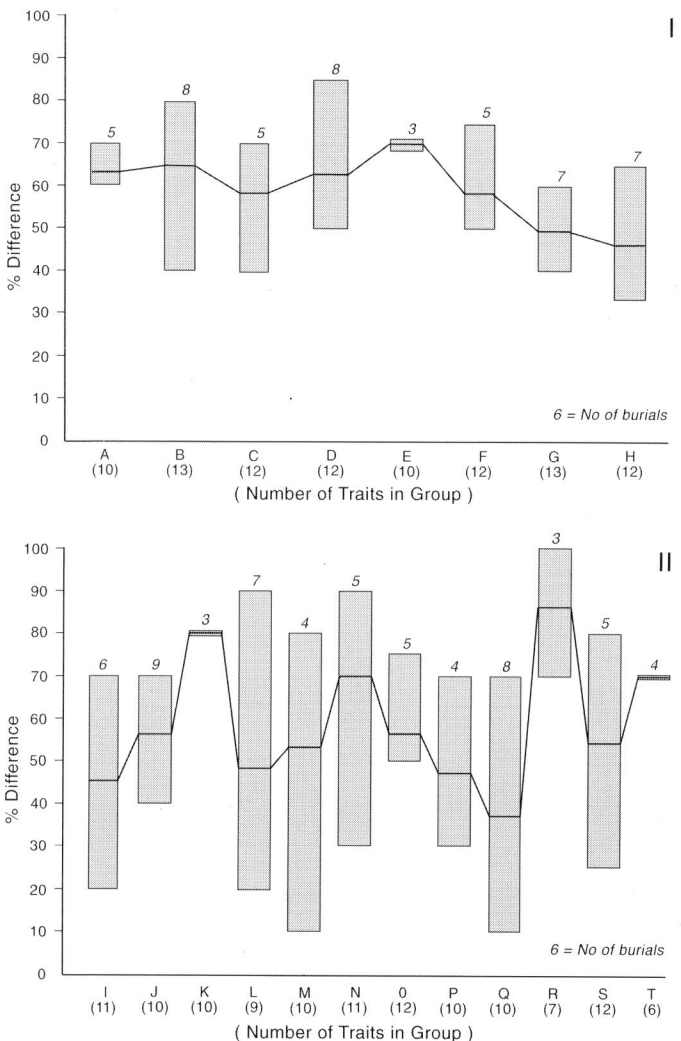

Fig. 30. Range of contrasts in mortuary practices by Group, Tiszapolgár-Basatanya. I: Period I; II - Period II.

against which to create a new identity in the Early Copper Age. The 'local' rules of burial rite were developed in the first decades of the cemetery's use-life, gradually being reproduced as the cultural memory of appropriate forms of behaviour. Thus, over decades, the 'global' rules of the cemetery became consolidated with the increasing number of burials and through whatever convergence occurred in the associations between specific age-sex categories of people and categories of objects and burial rites. However, the development and consolidation of a set of 'global' mortuary rules was in direct contrast to the principle of difference established, as well, in the earliest grave Groups at Basatanya. The playing out of this tension at funerals became the metaphor for the tension between structure and agency which is a key theme in this book.

The traits characterising groups, age-sex categories and other categories in Period I can be compared and contrasted with the set of 'global' mortuary rules defined after the fact through conventional archaeological analysis. In fact, there is surprisingly little deviation from these global rules, with the exceptions of a few group traits defined by different body position, body orientation or grave form. Burial traits are sometimes used to differentiate sub-groups in grave lines; the most frequent is body position, characterising Group G females, males of different ages in Group A and females of different ages in Group D. The co-variation of the quantity of deposited vessels and age is not found in Groups A and C but is otherwise strong. Sexual differentiation by unworked (female) versus worked (male) stone occurs but rarely (Groups D and G), with this trait used to distinguish younger from older individuals in Group C. Otherwise, the global rules developed through repeated social practice by the Basatanya community as a whole work reasonably clearly as a structure for mortuary behaviour. Other means, which rarely cut across community practices, have been found for the differentiation of graves in their lines. The same item of material culture is often used to signal a different message in different grave lines. It is important to note that some Period I grave lines hardly deviate from the global rules, while other lines can be strongly differentiated. The ratio of grave lines with no or one deviation to two or more deviations from the global rules in Period I is 1:1.6.

In Period II, a higher proportion of grave lines than before is characterised by traits in concurrence with the Basatanya global rules, in particular grave orientation and body position. However, burial traits are also widely used to differentiate sub-groups within grave lines. While grave form and depth are also commonly used, body position is used to define males of different ages in Group R and different individuals in Groups I, J and K. In addition, group traits varying from the global rules include left-sidedness (Group I), right-sidedness (J), irregular grave form (M) and non-standard orientation (O, S and T).

The overall pattern in Period II is that burial rites are used more intensively than before either to reinforce grave line identity with the overall community or to differentiate the line from the community. With grave goods, most grave lines support the global rules, while other lines choose neither to support, nor to deviate from, such rules. However, contrary to the global rule concerning sex-related pottery, vessel frequency in many burial lines is age-sex related (Groups K, L, N, O, and R), age-related (J, L, S) or individual-related (I, J, M, Q and T). While commentators from Bognár-Kutzián onwards have emphasised the decline in rule-boundedness in Period II, some grave lines show a weak tendency to deviation from community norms, while others are much more differentiated. The ratio of lines with no or one deviations to two or more deviations from the global rules is 1:3, rather higher than in Period I.

The interpretation of these data is difficult but one possibility is that concurrence with, or deviations from, global community rules is an indication of the tensions between the mortuary and the domestic domains. Overall rules develop from cumulative, small-scale, local patterns, which occur to a greater or a lesser extent in each of the grave lines. But distinctive mortuary practices in any single grave line manifest not only differences in identity between the deceased in that line but also a difference between that line and the whole community. If the assumption is accepted that each grave line derives from a separate, but related, residential group, it may be argued that similarities with the communal rules betoken an emphasis on relatedness and group cohesion, while differences indicate the importance of separate identity for the extended families of each farmstead. The ultimate in difference is the collapse of the old cemetery and the establishment of a new cemetery, in which burials are received from a new network of farmsteads. The greater emphasis on group cohesion in Period I may be related to the emergence of a stable pattern of community burial at Basatanya. The growing differences between grave lines in Period II may be explained in two ways. First, restriction of attendance at funerals to the immediate lineage of the newly-dead and/or to closely related lineages may well have produced greater variability in grave rites and material culture. Alternatively, increasing close ties to other social networks may have occurred with the passage of time, with the new groups manifesting their own identities in the form of different material culture or burial rites. It is difficult to adjudicate between these two potential explanations at present.

4.5 Summary

The Tiszapolgár-Basatanya cemetery marks a special place in the local landscape – a mortuary domain serving a network of small hamlets and scattered farmsteads which, in themselves, have diverse connections to wider social networks in the northern Alföld and beyond. It seems highly probable that the Basatanya cemetery formed the spatial framework for burials of closely-related members of extended kinship networks, while the grave lines found in both Periods contained members of more focussed kinship groupings, perhaps members of the same co-residential unit of dispersed farmsteads. The main sources of variability at Basatanya have been defined in terms of five forms of identity – group, age/sex, age-based, sex-based and individual. Group identities were clearly signalled through both burial ritual and grave good deposition; their importance increased in Period II. But, as a counterbalance to group identity, categorical traits, emphasising age and/or sex-based identities, were given special prominence in both Periods. Sex-based identities were of particular importance in Period I, while age/sex-based identities increased in importance in Period II. Although individual identities were always of some importance in some graves in each Group, individual identities increased in significance in Period II. In this same period, the deviations of local grave lines from the global rules structuring Basatanya either increased, to produce strong identities for several grave Groups, or declined, to produce Groups with strong identification with the overall community.

5. The Late Copper Age cemetery at Budakalász

5.1 Introduction

The major changes in cemetery nucleation and settlement dispersion which began in the Early-Middle Copper Age continue in the Late Copper Age in most areas of Eastern and Central Europe – not least in Hungary. In these post-climax centuries, settlements are so small and discard so minimal that it has often been difficult to locate sites. Those few tells still occupied in the Middle Copper Age are abandoned and settlements become dispersed into what appear to be small farmsteads, which either exhibit a degree of mobility or adopt radically different attitudes to material culture or conceivably both. The main landscape features are large cemeteries of the type found in Western Hungary and the Danube valley: few such large cemeteries are known from the Plain. The best examples of such cemeteries is Budakalász, with over 400 graves. A sample of more than a quarter of the graves from this site has been published in detail (Soproni 1956; Banner 1956) and will form the basis for a mortuary analysis of this period.

Baden cemeteries are best known for two related innovations – animal burials, usually of cattle, and fired clay cart models. The inclusion of these two classic features of Sherratt's (1979) secondary products revolution (or 2PR) – the interlinked suite of innovations shifting the balance from primary to secondary animal products – in the mortuary domain for the first time has been taken by Sherratt to mean that the Baden period is the time when the 2PR took off. But there is convincing evidence that the 2PR was less unified and that some elements of the so-called 'package' began rather earlier in different parts of SE Europe than others (Chapman 1982). More recently, Whittle has analysed the quartet of bodies – two bovine, two human – in grave 3 at the Baden cemetery of Alsonemedi, located SE of Budapest like Budakalász, in terms of new relations between animals and humans, including possession, fertility and regeneration (Whittle 1996:122-5). Most importantly, the humans in the graves were defined in relation to the cattle and vice versa, in a way which treated animals very differently from in the earlier Copper Age, where body parts were placed in the grave as food tokens, *pars pro toto*. One may expect to find differences in the way that other humans were themselves treated in cemeteries of this period.

5.2 The Budakalász cemetery: background and previous analyses

The Budakalász cemetery lies on a low sand hill some 0.5 km from the river Danube, N of Budapest (Soproni 1956: 111-128; Banner 1956: 188-208). Burial of the site under Danube sand has led to good preservation of the deposits. The graves consisted mostly of contracted inhumations, with a small number of extended inhumations and several 'cremations'. These cremations are not, however, deposits of the burnt, if not calcined, remains of disarticulated or destroyed bodies; rather, inhumations have been buried and a large fire lit in the grave (e.g., grave 113, where the charred upper bones of two bodies lay in an 8-10 cm layer of fine ash: Soproni 1956:126). Given the good preservation of the bones, it is disappointing that Nemeskeri has not identified the skeletons with any more specificity than as adult male, adult female and child, with no attempt at ageing the adults. This minimal categorisation limits the analysis but does not prevent the recognition of strong patterning in the cemetery.

Since there have been no major prior attempts at analysis of the published mortuary sample from Budakalász, it may be useful to outline the main dimensions of variability in the cemetery before proceeding to an analysis of the variations by grave line. The plan of the 115 published graves (Soproni 1956: Abb. 28; here = *Fig. 31*) shows signs of linear planning, combined with a number of outlying graves. The empty spaces between grave groups suggests that three zones can be distinguished: – the West zone, the North zone and the South East zone. The grave groups are not so clearly defined as the regular lines of graves at Basatanya – indeed, others may perhaps propose other groupings. The implied spatial ambiguity is typical of other aspects of this Baden cemetery – fuzzy sets rather than rigid dichotomies are the norm.

The proposed groups have been considered in terms of five variables: grave form, grave orientation, number in graves, age-gender categories and grave goods (Chapman 2000). Analysis of grave forms tends to show the predominance of oval over rectangular over other shapes. Grave orientation is variable; there is little regularity about the orientation of the graves at Budakalász, in contrast to both Kisköre and Basatanya. Fourteen out of the 16 possible compass points were utilised, with the only preference shown for SSE, S and SSW (half the graves). As Banner correctly noted (1956: 189), there was no clear correlation between orientation and age-gender categories, any more than there is between age-gender classes and the form of the graves- rectangular, oval or round. Each grave form is used for burials of adult women, men and children. The only distinction is that children are six times more likely to be buried in round graves than are adults.

In strong contrast to Basatanya, there is no age-gender regularity in the direction that the skeletons were facing. In single graves, there was a slight preference for right-sidedness among females and a stronger preference for left-sidedness among males but all placings were used for each age-sex category. However, a different message is conveyed in the multiple burials, with their greater potential for emphasising polarities and relationships. The patterning in the fifteen double and the single triple graves is that adult females were never placed on their right sides while more adult females than males were placed on their left sides. Conversely, children used all possible right-left combinations, while the 'Other' category (for extended burials and cremations) was filled by adults only but males and females.

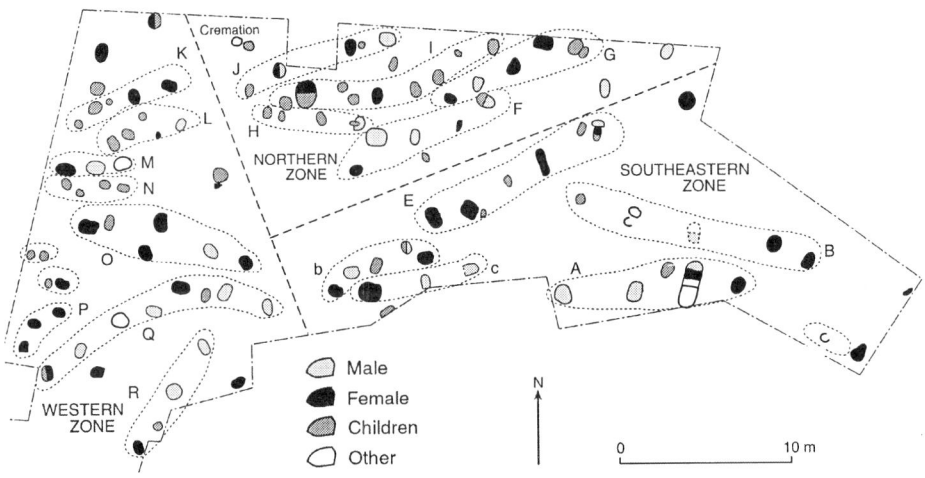

Fig. 31. Plan of published graves, Budakalász cemetery (source: Soproni 1956: Abb. 28).

There are three classes of grave offerings at Budakalász – ornaments which are regularly combined into forms of mortuary costume, vessels which are deposited individually or combined into drinking sets and other tools and raw materials. The ornaments are made of four materials – shell, polished stone, copper and (once!) animal bone. The individual items are combined into a variety of sets to form nine elements of costume, placed in different zones of the body: – head ornaments, necklaces, appliques on the chest, shoulder and back, bracelets, belts, knee-rings and ankle-rings. Of these nine elements, by far the commonest and also the most complex is the necklace, with the 38 examples from the cemetery comprising 21 different raw material combinations.

A similar pattern is found in other costume elements, with not one single raw material combination adorning all classes of body.

The commonest ceramics found in the Budakalász graves form the basic Baden drinking set – dippers, cups and jugs (*Fig. 32*), with bowls far less common and unmatched 'orphan' sherds also regularly deposited. Vessels were regularly placed in all but one of the six body zones – the exception was the torso, the main zone for costume display. The complete drinking set is deposited in only one grave – the adult female Grave 75, where two complete jugs and two complete cups were accompanied by fragments of a jug, a dipper and two more cups.

Fig. 32. Ceramic drinking set, Budakalász: 1: dipper; 2: cup; 3: jug.

The 'Other' category comprises bone, antler, teeth and stone as worked or unworked materials or artifacts, as well as copper. A special grave excluded from general discussion is Grave 91, interpreted by Banner (1956: 195-6) as a flint-knapper's grave because of the bag of variegated lithics found on his thigh. This adult male held jasper flakes in each hand, while an antler hoe (perhaps a status axe) was placed on the left shoulder-blade and flint and quartz flakes and chalk lumps placed near the pelvis. In the bag were two miniature serpentine axes, two bone awls and a bone spatula head, two caprine bones and 14 wild boar incisors, a lump of haematite, 6 red stones, obsidian microblades, flakes made of limestone, volcanic tuff, chalcedony, flint, jasper and obsidian, and debitage of flint, hornstone and jasper. No other grave contains such a wide range of raw materials; indeed, this is the only grave with obsidian, tuff, haematite, chalcedony and serpentine axes. In view of the other elements of the agrios (Hodder 1990) not connected to flint-knapping (the boar tusks, the axes and the haematite), it may be proposed that this is the grave of a long-distance specialist – perhaps a shaman – whose task it was to connect the local community to the wider world of the Late Copper Age.

Aside from Grave 91, worked and unworked objects occur in all body zones, with all lithics placed in the pelvic area. The widest variety of materials is placed near the skull, followed by the feet. While there are 11 graves with combinations of costume elements, or pottery, or 'Other' objects in different body zones, the most powerful demonstration of personal identity is the combination of different classes of grave goods on differing body zones.

These patterns may be summarised in the form of nine global rules for the mortuary practices of the published sample from the Budakalász cemetery:

1) The graves were laid out in loose linear groupings of unequal size.
2) The predominant form of burial was individual contracted inhumation, but some other modes of burial were practised, including multiple burials of humans and even animals.
3) The grave form was variable, with a predominance of oval over rectangular over other forms, but with no correlation with age-sex categories.
4) There was a tendency for varied grave orientation, showing no correlation with age-sex categories, with the exception of children and round graves.
5) There was no correlation between the sidedness of bodies in graves and age-sex categories. However, in multiple graves, there was a stronger, but still variable, patterning in sidedness in multiple graves.
6) There was complex and variable deposition of costume sets with all age-sex categories, most frequently placed on the torso.

7) There was variable deposition of ceramic drinking sets, or elements of these sets, with all age-sex categories.
8) There was variable deposition of other artifacts categories with all age-sex categories.
9) Both unworked and worked artifacts were placed in all body zones.

5.3 Micro-tradition analysis

We now turn to an analysis of the individual grave lines (*Fig. 31*). Because of the relative spatial disorder at Budakalász, it is difficult to apply the same rigorous criteria of group rectilinearity and grave spacing as at Basatanya. While some grave groups display strong rectilinearity (e.g., Groups E and K), others are notably curved (e.g., Groups D and Q). Nonetheless, even the curvilinear grave groups exhibit an integrity which argues for their validity as a group.

The criterion of grave spacing used is that two graves less than 5 m apart could be included in the same group. In the vast majority of cases, the inter-grave spacing is 3 m or less. Hence, grave 37 is excluded from Group B since, although sharing the same orientation, it lies over 8 m East of grave 18. Likewise, triple grave 47 is excluded from Group N because it lies 6.5 m East of grave 62. An exception is made in the case of Group J, where a gap of just over 5 m had been created by the destruction of the terrain midway along a strongly rectilinear Group.

The final criterion is more simple to apply; a Group must consist of a minimum of three graves. Thus, pairs of graves such as 85 and 86, 65 and 87, or 16 and 26, are excluded from the analysis. Occasionally, these criteria are combined to justify an exclusion. Such is the case with the arc of graves at the NW corner of the cemetery-graves 102, 103, 104 and 106, where there is a gap of over 6 m between two pairs of graves. Another feature of Budakalász is that there are several cases where a single grave is found off the line of the main Group (e.g., grave 7, near Group C; grave 42, near Group F; grave 60, near Group Q). It is possible that these graves were later additions to an already established grave line, whose positions bore a special relationship to particular graves in the middle of a line.

The application of these criteria means that 18 groups of burial lines can be postulated: five in the South Eastern zone (A-E), five in the Northern zone (F-J) and eight in the Western zone (K-R). Eighteen graves (15%) are not placed in groups.

The first problem faced by any attempt at analysis of grave lines is that there is no evidence for the point of origin for the cemetery. There is no published evidence for the occurrence of any preceding period of use of this site (Soproni 1956; Banner 1956). The subsequent analyses are based upon three assumptions: – (1) all of the grave lines start from a single, common direction; (2) the direction of origin is conventionally placed as East; and (3) since there is no clear evidence for the sequence of grave lines in the cemetery, the conventional starting point is the SE corner. However, each of these assumptions may readily be questioned; the spatial variability at the Kisköre settlement (see chapter 3) indicates that not all mortuary sites show such strong spatial order as the Basatanya cemetery (see chapter 4). Since excavations revealed that the cemetery of some 400 graves extended beyond the area shown in Soproni's plan on all sides, the excavated sample discussed here forms but one part of a much larger entity (p.c., P. Raczky and A. Endrődi).

An important point is that, in the only two instances of inter-cutting graves, the earlier of the two graves lies to the East of the later grave! Such limited stratigraphic data support the notion of a starting-point in the East, especially since both cases lie at the Eastern end of their respective lines (grave 51, cut by grave 50, in Group K; grave 98, cut by grave 97, in Group L). There is no *a priori* reason for a NW-SE sequence of grave lines; even though the stratigraphic evidence is limited, it is suggestive that the cemetery may have been laid out according to another spatial order, starting in the SE corner.

If the gaps between the three zones are taken to mean that burial in each zone was relatively independent of burials in other zones, it is possible that each zone used a different cardinal point for its point of origin, with grave lines oriented towards the common centre of the cemetery. However, there is another spatial logic, namely the unfolding of an integrated cemetery layout, with the grave Groups in the Western zone following a sequence from North to South, beginning from the end of Group J, the Northernmost and therefore the last in the Northern zone. At present, it is hard to evaluate the two possible arrangements. Neither do we have any information on the temporal spacing between burials or the likelihood of coeval burials in different lines. The complete absence of radiocarbon dating for Budakalász remains a serious research problem, which merits urgent action, especially since there is currently no means of locating the cemetery within the overall duration of the Baden period of approximately 800 years, or cca. 3600-2800 CAL BC (Forenbaher 1993: 246, Fig. 4).

The contents of the burials in the resulting 18 grave Groups are as follows (Table 34):

Table 34. Age/sex composition of grave lines and other graves, Budakalász.

Group	Adult Female	Adult Male	Child	Multiple	Incomplete Information	Cenotaph	Total Bodies (Graves)
A	2	1	1	1*6	-	-	6 (5)
B	-	2	1	-	1	1	5 (5)
C	2	1	-	-	-	-	3 (3)
D	1	1	2	1*5	-	-	6 (5)
E	-	4	4	-	-	-	8 (8)
F	1	2	-	2*4&5	-	-	7 (5)
G	1	3	2	-	-	-	6 (6)
H	1	-	4	-	-	-	5 (5)
I	-	1	6	2*1&3	-	-	11 (9)
J	1	1	2	1*3	-	-	6 (5)
K	-	2	2	1*1	-	-	6 (5)
L	1	1	2	1*1	-	-	6 (5)
M	1	1	-	-	1	-	3 (3)
N	-	-	2	1*1	-	-	4 (3)
O	1	4	1	-	-	-	6 (6)
P	-	3	-	-	-	-	3 (3)
Q	4	1	1	1*2	1	-	9 (8)
R	2	1	1	-	-	-	4 (4)
Total/*A Bodies	23	35	42	-	3	1	104
Total/ Graves	18	29	31	11	3	1	93

Key: *1 – 2 children; *2 – adult male + child; *3 – adult male + adult female; *4 – 2 adult females; *5 – adult female + child; *6 – adult female, adult male + two cattle. *A – totals include bodies in multiple burials counted as adult males, adult females or children.

The overall distribution of age-sex categories indicates a drastic difference from that of the Basatanya cemetery. Here, the largest single category comprises children, outnumbering by 20% the adult males, who, in turn, outnumber the adult females by 50%. Hence, there are several grave lines in which children outnumber adults and one Group in which only children are interred. Only a single cenotaph grave is included in the grave Groups. A second cenotaph (grave 26) lies apart from the grave lines and, if the hypothesis of a SE starting-point for the cemetery as a whole proves correct, could be a very early deposition. But, as at Kisköre and Basatanya, the main impression from the age-sex composition of the Budakalász Groups is one of great variability, with no single Group replicating the contents of any other Group. The number of bodies varies from three to 11, with the number of graves varying from three to nine. The main difference from Basatanya and Kisköre lies in the number of multiple graves, which provides a new means of measuring the difference between pairs of burials.

Group A (Fig. 33 & Table 35) The first grave line is a small Group, with all age-sex categories represented but with more adult males than females. The line is spatially dominated by the largest grave in the excavated sample – the quadruple grave of two adults and two cattle. Here, the cattle are inhumed in the same way as the humans, but placed upon their stomachs – a position never used for humans at Budakalász. No grave goods are placed with the cattle but the contrasts between the adult female and male are emphasised by different orientation and by minor differences in objects.

The only Group trait common to all graves, including the cattle burials in grave 3, is contracted inhumation. Both grave form and grave depth exhibit a moderately correlated wave pattern of contrasts between each pair of burials: oval graves tend to be shallower, and rectangular graves deeper, with the exception of the Westernmost grave 2 (oval and deep). This variability is emphasised by the grave orientation, which differs for each pair of graves but is unrelated to the variations in depth and form. It is also enhanced by contrasts in body position between most pairs of burials, similar for only one pair of successive interments. Thus, almost all burial rites underline the variability between persons successively buried in this Group, variations which occur irrespective of age-sex similarities or differences in each successive pair of burials.

The scale of contrasts between successive burials in this group is moderate, ranging from 25 to 60% on a total of 15 traits. The smallest contrasts occur between the paired adult female and adult male burials in quadruple grave 3, while the greatest occur between two adult females in the last two graves.

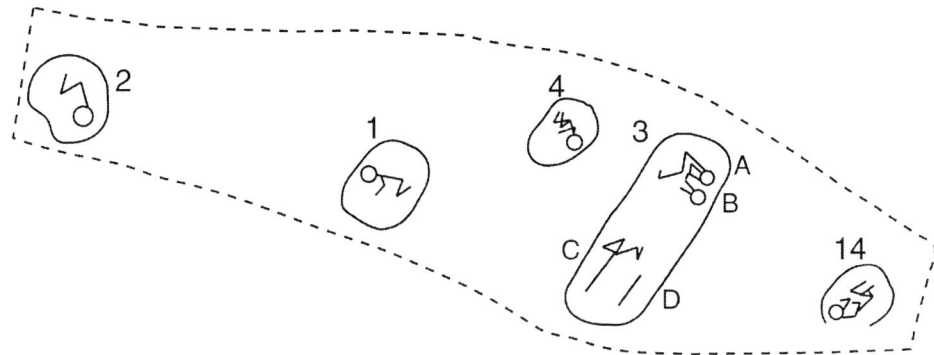

Fig. 33. Group A, Budakalász.

Table 35. Grave good, costume and animal bone contrasts in Group A, Budakalász.

Grave 2	Grave 1	Grave 4	Grave 3	Grave 14
bead necklace	-	-	-	-
dipper	dippers	-	cup	body sherds
sherds	-	-	-	-
-	animal tooth	-	-	-
-	-	-	copper awl	-
flint flakes	-	unworked stone	flint flakes	-

The limitations of ageing and sexing of the excavated sample of skeletons means that the only definable age-based traits indicate contrasts between children and adults, with no possible age contrasts between adults of different ages or indeed infants and adolescents. Likewise, the only recognisable sex-related traits are based upon differences between adult males and females, with children being excluded. Only one sex-based trait can be identified in this Group: the only burials with complete vessels are women's graves, while sherds or no pots at all are found with children and adult male graves. The only age-based trait is based on the discovery in the only child's grave of large, unworked stones in the grave fill; in contrast, no adult graves include such material in the fill. The discovery at Budakalász of many graves with stones of varying sizes, sherds and lithics placed in the grave fill makes this a significant mortuary practice and not just a form of accidental incorporation, as has been argued for modern Polish graves by Buko (1998). All the other traits representing difference may be related to individual contrasts, whether between the adults in the quadruple grave (the presence versus the absence of a copper awl, a cup or flint flakes), between two females (different locations for similar dippers) or between all females (presence or absence of (a) a necklace of copper, Dentalium and shell beads, (b) flint flakes or (c) animal teeth).

Group A exhibits a surprising degree of variation between successive graves, based largely upon contrasts in burial ritual but also upon individual traits rather than on age-based or sex-based traits. The two adults buried with (their) cattle in grave 3 are the pair showing the least mortuary variability of any pair in the Group.

Group B (Fig. 34 & Table 36) Group B consists of a small number of widely-spaced graves, with more adult males than children. There are no multiple graves but a single example of a cenotaph. Apart from the cenotaph, all the other graves are contracted inhumations, signifying a Group trait. The other possible Group trait is a grave orientation from SSE-NNW, found in the three graves where information is available. Grave orientation at Budakalász is usually so variable that the coincidence of three graves dug in the same direction is so rare that it should be counted as a Group trait. The grave depth in Group B follows a wave pattern, much as in Group A, with the first grave dug shallow and alternating deeper and shallower graves in sequence.

The scale of contrasts in this Group is high, with a range from 45 to 70% based upon nine traits. The least contrasts are found between two pairs of graves – two adult males and an adult and the cenotaph; the strongest contrasts were found between the cenotaph and the child's grave at the W end.

Table 36. Grave good, costume and animal bone contrasts in Group B, Budakalász.

Grave 30	Grave 31	Grave 15	Grave 19	Grave 18
-	-	-	shell applique	-
painted cup	-	-	-	-
-	-	-	dipper	dipper
			beakers	jug
-	unworked stones	-	-	-

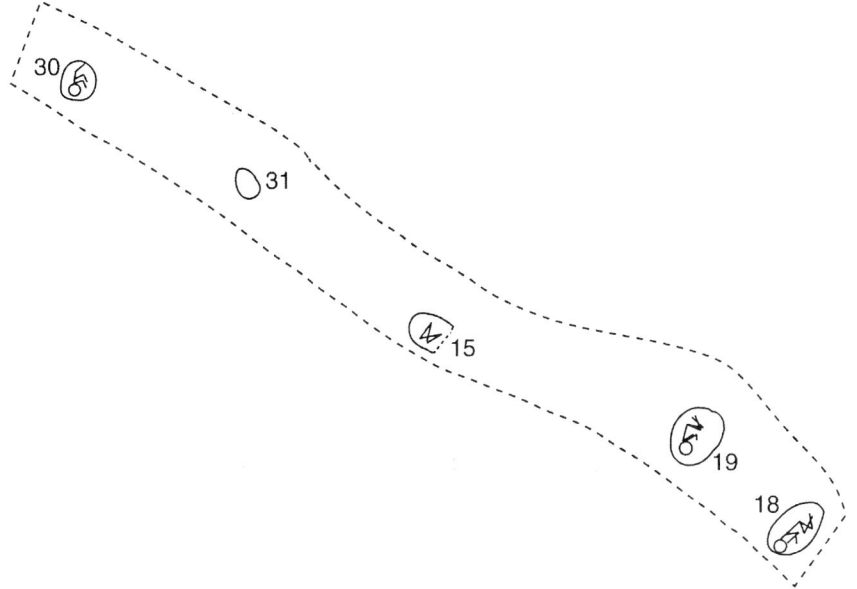

Fig. 34. Group B, Budakalász.

Since the graves in Group B are not richly furnished with grave goods, there are relatively limited possibilities for differentiation. The only categorical contrast is that unworked stones are found in the fill of the cenotaph, rather than in the child's grave, as in Group A. Other contrasts are found as individual traits, between adult males, as with the presence / absence of shell applique costume. The deposition of different vessels in the two adult male graves shows a further contrast from yet a different vessel in the child's grave. No sex-based or age-based traits were utilised in this Group.

Group C (Fig. 35 & Table 37) This small Group of widely-spaced graves comprises two adult female and one adult male grave, with no children or special graves. There are contrasts between the second and third graves in all aspects of burial rite except grave orientation, which shows maximum variability. However, the other contrasts never overlap, leaving a complex pattern of variation between individuals. One of the female graves is a rare example of an extended inhumation on the back – an individual trait difference from that of the other adult female. In small Groups, the scale of contrasts is usually high and this is the case here, with a range from 45 to 80% based upon nine traits.

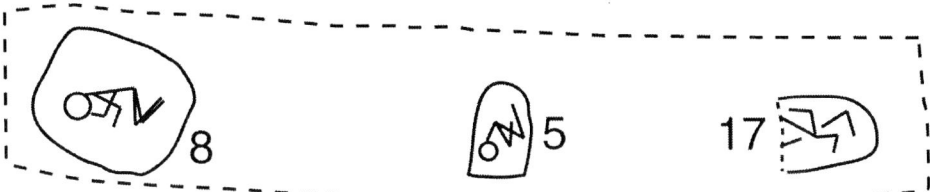

Fig. 35. Group C, Budakalász.

Table 37. Grave good, costume and animal bone contrasts in Group C, Budakalász.

	Grave 8	Grave 5	Grave 17
necklaces		-	-
cup		-	-
worked lithics		unworked stones	-

The contrasts between the male and one female are stronger than between the two females. All of the grave good traits in this Group are found with the sole adult male burial, thus reinforcing sex-based differentiation. The only individual contrast between the two female graves is the presence of unworked stones in the fill of one of the graves.

Group D (Fig. 36 & Table 38) The larger Group D lies close to, and parallel with, Group C. Group D comprises five burials, the second of which is double. The Group composition is balanced for adult females, males and children. There is a strong group identity in this Group, with three Group traits in the burial rites. All the burials are contracted inhumations, while four out of five graves are oval and five out of six bodies have been laid on their right sides. Even the grave depths form a tight grouping, with all graves except one (the Westernmost grave 9) within 10 cm of each other. Only in grave orientation are the contrasts between each pair of burials fully exploited. The overall scale of contrasts in this Group is thus low, ranging from 25 to 55%, based upon 11 traits. The smallest contrasts fell between two pairs of burials: an adult female and a child in the double grave 13, and two children.

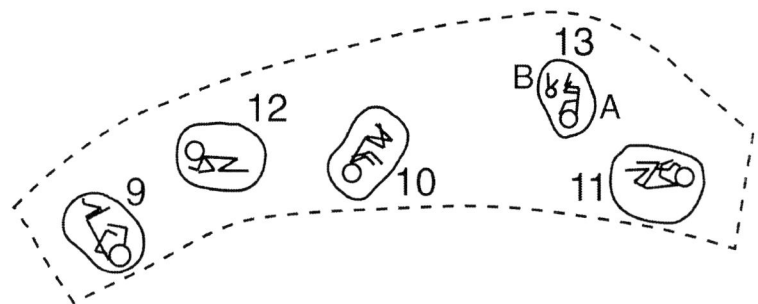

Fig. 36. Group D, Budakalász.

Table 38. Grave good, costume and animal bone contrasts in Group D, Budakalász.

Grave 9	Grave 12	Grave 10	Grave 13	Grave 11
-	shell bracelet	shell necklace	shell necklace	-
			shell bead belt	
cup	cup	-	-	-
-	red stone	-	-	-

The paired burial shows other contrasts, too: only the female has grave goods, a shell plaque necklace and a shell bead belt. These costume elements contrast with a shell and shell bead necklace found with another child and a shell, shell bead and shell disc bracelet found on the arm of the other female. While these costume elements constitute individual traits, they also form a sex-based trait since no costumes are found on the sole adult male. The deposition of cups near the skulls of an adult male and an adult female grave indicate individual differences between males, and between females, while linking the two graves at the W end of the line. Finally, the presence of a red stone in only one female grave comprises another female trait. Thus, as in other Groups at Budakalász, Individual traits dominate the domain of grave good deposition, while Group identity is more regularly defined in burial practices.

Group E (Fig. 37 & Table 39) The individuals of this large Group are evenly split between children and adults, with three out of the four adults specified as males and the other an unsexed adult. The seeming absence of adult females in such a large Group is surprising; neither are there multiple graves nor cenotaphs in this Group. The children are located in the middle of the Group, as is the case in many other grave lines. Only one burial trait is consistent enough to be considered as a Group trait – the contracted inhumations found in seven out of eight bodies. The remaining four burial traits all provide contrasts between most successive pairs of burials – body position, grave form, grave orientation and grave depth, which typically follows a wave pattern starting with a deeper grave at the E end. With the possible exception of oval graves whose orientations are mostly focussed on SE-SSW, these burial traits do not form a highly correlated suite of variations but rather form an overlapping set of contrasts.

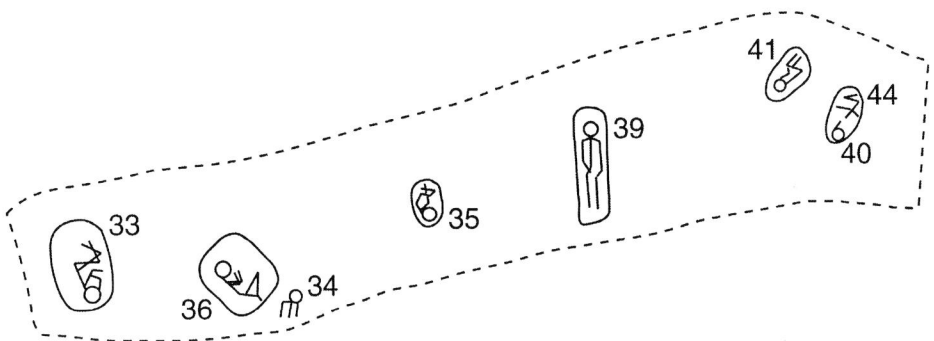

Fig. 37. Group E, Budakalász.

Table 39. Grave good, costume and animal bone contrasts in Group E, Budakalász.

Grave 33	Grave 36	Grave 34	Grave 35	Grave 39	Grave 41	Grave 40	Grave 44
shell belt	-	-	shell belt	-	-	-	-
-	necklace	-	-	-	necklace	-	-
jug, cup	jug, cup	-	-	body sherds	-	-	jug
unworked stones	-	-	unworked stones	-	-	-	jasper scraper
							polished red stone

The scale of contrasts is one of the widest for Budakalász, ranging from 20 to 85%, based upon 14 traits. The smallest range of contrasts is between a child and an adult male, while the greatest variations serve to different two adult males at the W end of the line. This is one of the very few Groups where maximum possible contrast scores are found for burial rites in two pairs of graves (adult male and child; two children). Apart from the two highly differentiated adult male graves, most other graves show fewer contrasts in grave goods than they do in burial rites.

No categorical differences are highlighted by the deposition of grave goods, which instead are related to individual differences within age-sex categories. With costume elements, similar shell belts are found with one child and one adult male, while different necklaces are deposited with another male and another child (copper, shell and white marble with the male; copper, *Dentalium*, shell, white and red marble with the child). Pottery is not deposited with children but the type of vessel and the body zone where it is deposited varies widely in adult male graves. Unworked stones are deposited in the fill of one adult male grave and one child's grave – the same graves as held the shell belts! The unsexed adult grave received a jasper scraper and a polished red stone-types not found in any other grave. Thus, this Group exhibits a high degree of individual variation in most burial rites and in a number of grave goods. The emphasis on individual traits is apparently at the expense of traits signifying Group identity.

Group F (Fig. 38 & Table 40) This medium-sized Group comprises five graves, two of which are double burials. The Group is dominated by adult females, with a minority of children and one single adult male. The first burial is a double grave with a child and adult female (grave 56); the last-but-one is a double grave with two adult females (grave 48). Two Group traits are found in the burial rites: contracted inhumation occurs in six out of seven cases, while oval graves are found in four out of five cases. For each trait, the exception emphasises a contrast related to a double grave: an extended inhumation on the back in the first double grave, a round grave in the more Westerly grave. As is often the case in Budakalász, the grave orientation establishes points of contrast in each successive pair of burials. Both body position and grave depth vary with the place of the grave: right-sidedness and shallower graves occur to the E end, left-sidedness and deeper graves to the W end of the line.

The scale of contrasts in Group F is on the low side, varying from 15 to 50%, based upon 12 traits. The level of contrast varies between the two paired burials: a moderate number of contrasts occurs in the first grave, while the smallest number of differences emphasises the unity of the second paired grave, which

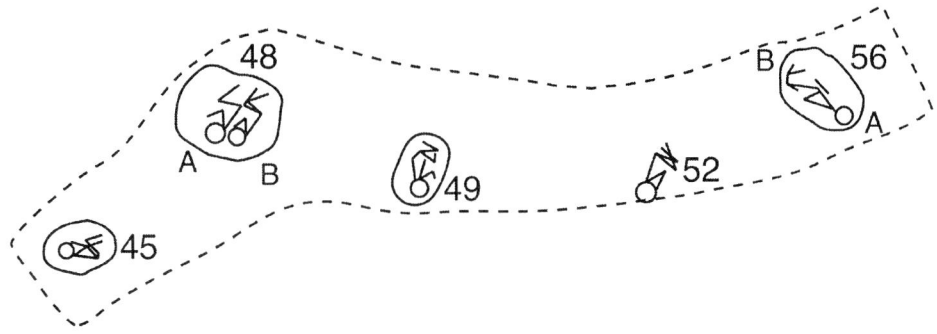

Fig. 38. Group F, Budakalász.

Table 40. Grave good, costume and animal bone contrasts in Group F, Budakalász.

Grave 45	Grave 48	Grave 49	Grave 52	Grave 56
-	bead necklace	-	-	shell applique
-	-	-	-	body sherds
unworked stones	lithic	jasper flake	-	-

includes identical burial rites. The greatest number of contrasts distinguishes an adult female from the only adult male.

The only categorical distinction found in this Group concerns the adult male grave, the only grave to contain unworked stones not only in the grave fill but also near the skull and the knee. Individual traits may be used to invoke all the other differences – most of which concern differences between paired burials. In grave 56, a shell bead applique adorns the child, while sherds are deposited near the skull of the female. In grave 48, a necklace of copper, copper/*Dentalium* and shell beads was placed round the neck of one of the females, while lithics were placed near her feet. The only other individual trait concerned the differentiation of one adult female from the others by the deposition of a jasper flake near the pelvis. Once again, Group F reveals a strong element of individual differentiation in the deposition of grave goods, with less variability in the burial rites.

Group G (Fig. 39 & Table 41) This Group is of similar size to Group F and lies parallel to it, adjacent to the North. The group consists of more adult males than children, with but one single adult female. All of the graves are single graves, with no special burials. Group G is one of the groups where there is a stratigraphic relationship between graves: the first, Easternmost grave (grave 97) is cut by the second grave (grave 98), which is some 15 cm shallower. Two Group traits are found in all the graves – contracted inhumation and an oval grave form. The grave depth conforms to a wave pattern, starting with a deeper grave. The main variability in the burial rites concerns grave orientation, which provides contrasts for most pairs of successive burials.

The scale of contrasts is low to moderate, ranging from 20 to 55%, based upon nine traits. Two pairs of graves display the smallest contrast – a child and an adult male, and an adult male and the only adult female. The biggest contrasts are exhibited at the two pairs of graves at the ends of the line: the children at the E end and the adult female and an adult male at the W end.

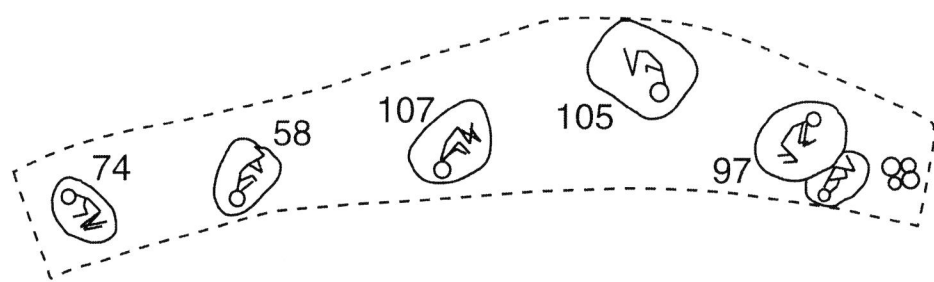

Fig. 39. Group G, Budakalász.

Table 41. Grave good, costume and animal bone contrasts in Group G, Budakalász.

Grave 74	Grave 58	Grave 107	Grave 105	Grave 97	Grave 98
shell necklaces	bead headdress	-	-	isolated shell	-
dipper	-	-	-	-	-
-		-	unworked stones	-	unworked stones

The only categorical trait found in this Group concerns the only adult female – a shell bead head-dress costume element placed on the skull, and found in no other grave. Otherwise, individual traits are used to differentiate both adult males and children. The only vessel deposited in this group – a dipper placed near the skull – is found with one adult male, who is also adorned with two shell and shell bead necklaces. Likewise, a single shell is placed on the neck of one of the children. Unworked stones are deposited in the fill of two graves – one of the children and one of the adult males. As in Groups E and F, Group G is dominated by individual traits rather than categorical traits, with rather few Group traits of identity either.

Group H (Fig. 40 & Table 42) This small Group begins at the very edge of Group F; the first grave is less than 2 m from double grave 48 (Group F). Nonetheless, the graves are placed in a convincing line; four of them are single graves, while the last grave, at the W end, is a double children's grave. The first grave (grave 51: depth – 130 cm) is cut by the much shallower second grave (grave 50: depth – 55 cm) – the second example of stratigraphic superimposition of graves in the Northern zone. It is interesting that the Group is dominated by children's burials in five out of six cases, the other burial being of a an adult female. Two Group traits can be observed: contracted inhumation and an oval grave form, both traits found in all burials. After the second, very shallow, grave, there is a linear decrease in grave depth, which would presumably have been a general trait had it not been for the cutting of grave 51. As usual, grave orientation is much more varied, with contrasts between most grave pairs.

The scale of contrasts in this Group covers a wide range, from 0% to 70%, based upon 16 traits. This is the only case in the excavated sample that two graves – children's graves 54 and 55B – are identical in all aspects of the rite, albeit neither burial has any grave goods. The greatest contrasts also occur between a pair of children.

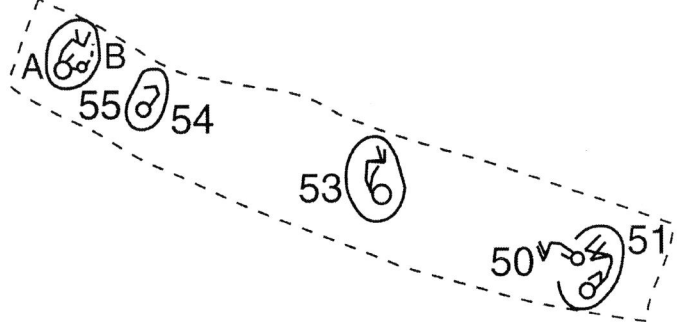

Fig. 40. Group H, Budakalász.

Table 42. Grave good, costume and animal bone contrasts in Group H, Budakalász.

Grave 55	Grave 54	Grave 53	Grave 51	Grave 50
shell necklace	-	shell applique	-	shell & stone bracelet
				shell applique
				shell bracelet
dipper	-	dipper	footed bowls	body sherds
-	-	jasper flake	-	unworked stone

The sole adult female grave is also the only grave with deposition of bowls near the skull and in the fill – indicating an age-based trait compared with the body sherds in one of the children's graves. All the other differences are accounted for by individual traits within children's graves. These include the differences between the two children in paired grave 55: burial 55A has a shell bead necklace and a dipper and small unworked stones placed near the knee (cf. a dipper near the pelvis of another child). Other children's individual differences include costume elements, such as the *Dentalium* applique on the scapula of one child, in contrast to the shell, copper and red stone necklace, shell applique on the torso and shell plaque bracelet of another child. Finally, one sole child's grave has a jasper flake deposited near the pelvis, while another has unworked stones placed in the fill. This mass of individual traits, some but by no means all related to the paired grave, indicates the significance of individual differentiation in this Group.

Group I (Fig. 41 & Table 43) This long, sinuous grave line is one of the largest Groups in the excavated sample of the cemetery. Nine graves contain eleven bodies, including two double graves at the W end of the line. These double graves characterise this Group as different from the others in the Northern zone, specifically because of the burial rite consisting of the lighting of a fire in the grave and the production of a burnt ash layer on which the burnt bones were found. As explained above, this does not constitute a cremation, but is rather a practice which seals the base of the grave and relates the body to the element of fire. All of the single graves contain contracted inhumations laid out in the traditional manner.

The Group resembles Group H in its dominance by children's graves: eight out of eleven bodies are children, with two adult males and a single adult female. The first double grave contains an adult female and an adult male; the second double grave contains two children. Although there are more round graves in this Group than is normal, this special grave form is not widespread enough to form a Group trait. Grave orientation clusters around the South point, but there are sufficient SSEs and ESEs to make this a variable practice, again insufficient for a Group trait. Grave depth does not strictly form a wave pattern but there is variability without a clear linear pattern. It is unusual to find a grave good trait which may constitute a Group trait. Here, in Group I, the negative trait of a complete absence of pottery deposition makes it such a candidate.

The scale of contrasts is very varied, ranging from 10 to 75%, based upon a set of 11 traits. The lowest set of contrasts occurs in the double grave 100, between an adult male and the only adult female. The greatest contrasts occur between two children's graves.

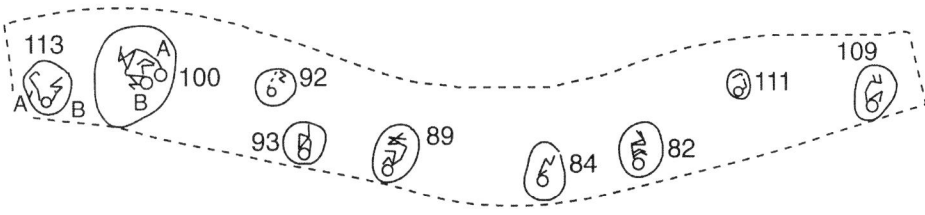

Fig. 41. Group I, Budakalász.

Table 43. Grave good, costume and animal bone contrasts in Group I, Budakalász.

Grave 113	Grave 100	Grave 92	Grave 93	Grave 89	Grave 84	Grave 82	Grave 111	Grave 109
applique	necklace	-	-	necklace	necklace	necklace	necklace	-
necklace				necklace			headdress	
-	-	-	unworked stones	-	jasper point	unworked stones	-	-

One possible sex-based trait occurs in Group I – the negative trait of the absence of costume elements in both of the adult male graves. This absence forms part of the contrast between the adult female and adult male bodies in double grave 100 – with a necklace of shell and shell beads placed on the female's torso. Otherwise, children's graves provide the context for most of the differences in individual traits: whether in the double grave 113, with its costume contrasts (copper bead necklace in one, copper and *Dentalium* beads in another necklace), or whether in single graves, with contrasts in costume and lithics. The costume contrasts concern a shell, *Ostrea* and red stone head-dress and three further variants on necklaces – a perforated shell plaque necklace plus a copper, *Dentalium*, stone and shell necklace on one and a *Dentalium*, red stone and shell bead necklace on another. Two children's graves contain unworked stones in the fill, while one grave has the deposition of jasper and hornstone flakes near the skull. Once again, individual traits dominate the variability of Group I, with negative traits relating to sex-based and Group identity.

Group J (Fig. 42 & Table 44) This medium-sized Group consist of a balanced set of all age-sex categories, in five single and one double grave (an adult female and an adult male). Group traits are restricted to two burial rites found in all cases – contracted inhumation in an oval grave – and a scarcely more variable third. Grave depth is so limited, to a range of 10 cm between 100 and 110 cm, that it may be considered as a Group trait. The main contrasts are formed by the grave orientation, which varies between each pair of burials. Because of these Group traits and the paucity of grave goods, there is a limited scale of contrasts, ranging from 25 to 40%, based upon 12 traits. The smallest contrasts are found in the double grave (including body position and grave orientation), the largest between an adult male and an adult female.

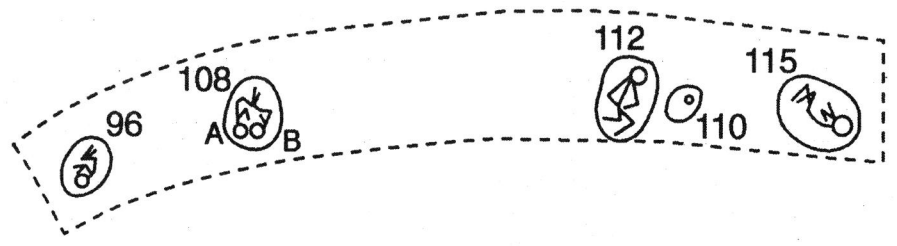

Fig. 42. Group J, Budakalász.

Table 44. Grave good, costume and animal bone contrasts in Group J, Budakalász.

Grave 96	Grave 108	Grave 112	Grave 110	Grave 115
shell necklace	-	-	-	-
-	-	jugs, dippers	-	-
-	bone awl	-	-	-
-	-	jasper flakes	-	jasper scraper
-	-	antler hoe	-	-

The differences between grave goods may be related solely to Individual traits, with no categorical differentiation in this Group. The sole item of costume is a shell necklace found with one of the children. Pots are found in one adult female grave (a dipper at the elbow) and one adult male grave (jugs and dippers near the pelvis) – with variations in both pot form and place of deposition. One aspect of the contrasts within the double grave is the placing of a bone awl near the skull of the female. Jasper tools are found with one female grave (a scraper near the hand) and one adult male (flakes near the pelvis and arm), an antler hoe being deposited with the same adult male. Mortuary variability in this Group appears to show a tension between Group traits in burial rites and individual variation in grave good deposition.

Group K (Fig. 43 & Table 45) This medium-sized Group is dominated by the burial of children, including two in a double grave, with adult males as well but not a single adult female. Group traits include contracted inhumation, in all cases with information, and probably the use of an oval grave, in four out of five cases. Grave orientation and depth show a parallel pattern of similar ends and contrasting middle burials – an unusual linear trend with contrasts between most pairs of graves. A wide range of contrasts is found, from 15 to 75% based upon a set of 12 traits. The smallest difference lies between an adult male and a child, the greatest between that same adult male and the male in grave 91, the potential 'shaman'.

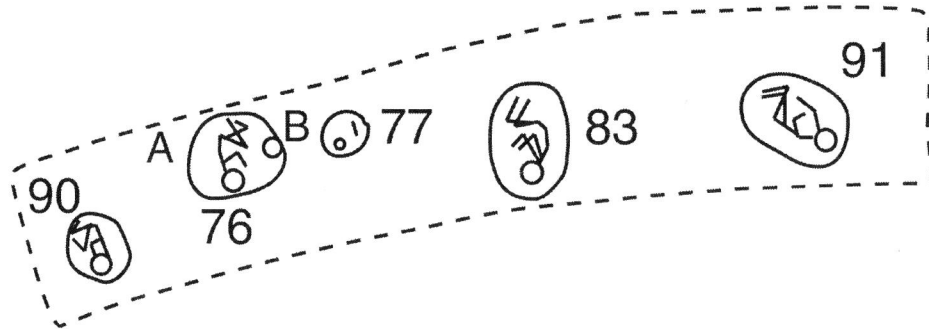

Fig. 43. Group K, Budakalász.

Table 45. Grave good, costume and animal bone contrasts in Group K, Budakalász.

Grave 90	Grave 76A	Grave 76B	Grave 77	Grave 83	Grave 91
shell necklace	copper frag.	-	-	-	-
unworked stone	-	-	-	-	jasper flake
	-	-			flint scraper
	-	-			bag of lithics
	-	-			chalk lumps
-	-	-	-	-	antler hoe

All of the differences in grave goods can be plausibly related to individual traits. The differentiation between the children in the double grave is minimal, consisting of a fragment of copper under the skull of one child. The cup and the unworked stones in the fill cannot be related to one child rather than the other. One other child's grave has unworked stones in the fill and a shell necklace was deposited in the same grave as the only item of costume. The variety of raw materials placed in grave 91 marks this male out as a special case (see above, p. 129). Again, as with Group J, variability in Group K refers to individual differences in grave goods combined with a strong element of Group identity in burial rites.

Group L (Fig. 44 & Table 46) This Group resembles Group K in both size and composition – dominated by child burials and with one double children's grave as well as one partial burial of a child at the W end. However, in addition to the children, there is one adult female as well as one adult male. Only one Group trait may be distinguished – the contracted inhumation rite. Grave depth appears to be age-related, with deeper graves for the adults and shallower for the children. Grave orientation presents contrasts with each successive pair of graves. The scale of contrasts in this Group is low, ranging from 20 to 45%, based upon 11 traits. Three graves pairs share the lower score (adult female and male; adult male and child; children in the double grave), with both pairs of children showing the higher score.

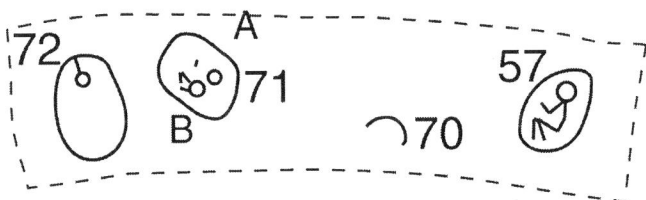

Fig. 44. Group L, Budakalász.

Table 46. Grave good, costume and animal bone contrasts in Group L, Budakalász.

Grave 72	Grave 71	Grave 88	Grave 70	Grave 57
perf. tooth	shell plaques	shell armband	-	-
	shell necklace			
-	body sherds	jug	footed cup	-
			jugs	
-	unworked stones	unworked stones	-	-

Two age-related traits differentiate the children from the adults – costume elements with all the children and neither adult, and unworked stones in grave fill of two of the children's graves (including the double grave) and neither adult grave. Each child has a different costume element, ranging from a single perforated animal tooth on the torso (presumably a pendant), to shell plaques on the torso, a shell bead necklace and a shell bead armband. The only individual differences refer to pottery, where two out of the three children's graves have body sherds placed under the skull, while a third has a jug placed in the grave fill. While no vessels were deposited in the female grave, the adult male grave has two jugs and the fragments of a footed cup in the grave fill. In this Group, age-related traits, in burial rite (grave depth) and in costume and lithics, assume greater importance than usual, with relatively few individual traits and a weak Group identity.

Group M (Fig. 45 & Table 47) This small Group consists of an adult female, an adult male and a disturbed grave, for which there is minimal anatomical information and no grave goods. Thus the only contrasts which can be identified relate to sex-based contrasts between the two adult graves. On the scale of contrasts, a 75% difference is observed.

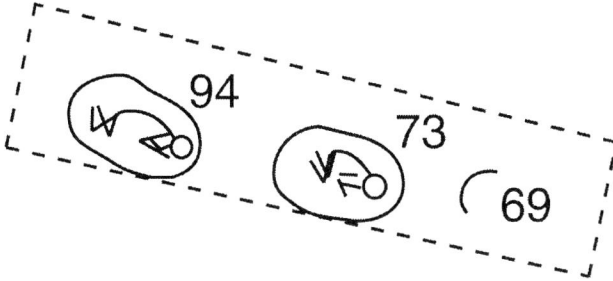

Fig. 45. Group M, Budakalász.

Table 47. Grave good, costume and animal bone contrasts in Group M, Budakalász.

Grave 94	Grave 73	Grave 69
shell necklace	shell & stone necklace	-
	bone plaques	
-	painted cup	-
	dipper	

All the burial rites are identical except for the grave orientation. Different types of necklace are worn, while, in addition, the female is adorned with bone plates on the torso and bone plaques on the leg. Two vessels are placed near the female's femur – a red painted cup and a dipper. The limited size of this Group precludes more general comments.

151

Group N (Fig. 46 & Table 48) This medium-sized Group consists entirely of children's graves (three single and one double) – the only such Group in the excavated sample. Only one Group trait can be distinguished – the contracted inhumation rite. In grave form, orientation and depth, there are contrasts between each successive pair of burials, with grave depth exhibiting a wave pattern, beginning with a shallow grave at the E end. The scale of contrasts is moderate, mainly because of the paucity of grave goods. The scores range from 25 to 60%, based upon a set of eight traits.

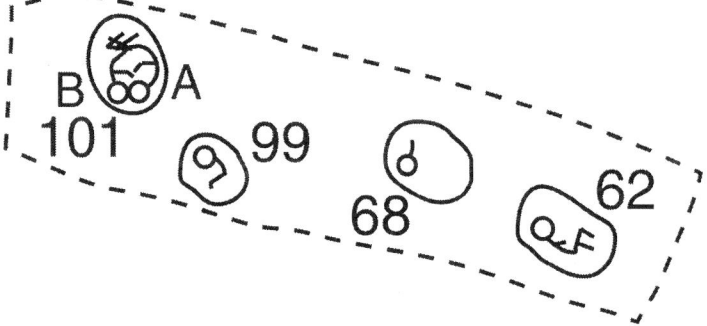

Fig. 46. Group N, Budakalász.

Table 48. Grave good, costume and animal bone contrasts in Group N, Budakalász.

Grave 101	Grave 99	Grave 68	Grave 62
4-footed cup	-	dipper, jug	-

The double grave reveals the lowest contrast score. The only grave goods are three vessels, two found with a single grave (dipper and jug near the skull), the remaining four-footed cup near the arm of one of the children in the double grave. By definition, all the differences in this Group relate to individual distinctions between children.

Group O (Fig. 47 & Table 49) This medium-sized Group, consisting of six single burials, is dominated by adult male burials, with one adult female and one child. There is a strong set of Group traits, including contracted inhumation in all cases, left-sided body position (five out of six cases) and oval grave form (five out of six cases). Grave depth is maintained within a relatively tight range of 20 cm (90-110 cm) but this is not close enough to qualify for a Group trait.

152

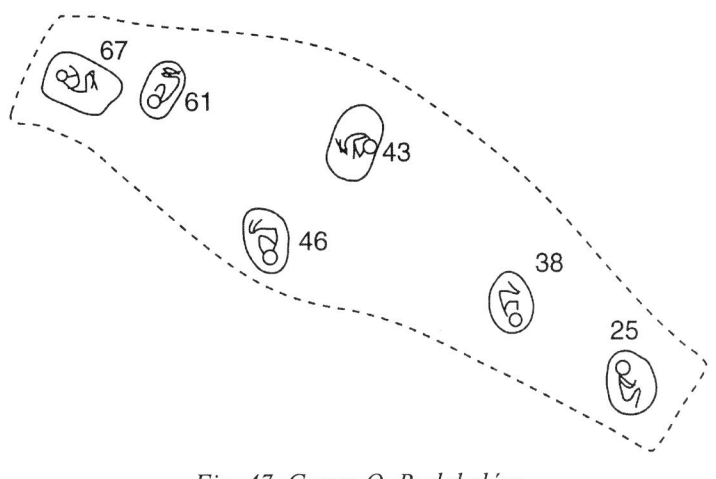

Fig. 47. Group O, Budakalász.

Table 49. Grave good, costume and animal bone contrasts in Group O,
Budakalász.

Grave 67	Grave 61	Grave 46	Grave 43	Grave 38	Grave 25
necklace	-	necklace	necklace	copper beads	necklace
-	painted cup	painted dipper	-	-	-
-	limestone flake	flint flake	-	-	red stone
-	unworked stone	unworked stones	-	-	unworked stones

The major source of variation is the grave orientation, which contrasts for each successive pair of graves. The scale of contrasts is moderate, with scores ranging from 35 to 60%, based upon 12 traits. The smallest score differentiates the only adult female from an adult male, the largest shared by two pairs: an adult male and the female, and an adult male and the only child.

Complex differentiation is exhibited in the costume elements found in all graves but that of the child – the only age-related trait. While the only female has copper beads attached to clothing on her torso (therefore a sex-based trait), all the males display individual traits of necklaces of shell and stone (but not copper), in three ways: (a) shell beads (1 case); (b) shell beads, *Dentalium* and perforated shells (2 cases); (c) shell beads, *Dentalium* and red and white marble beads (1 case). The only two vessels deposited in this Group are both red painted – one a cup in the only child's grave and the other a dipper in one of the male's graves – another individual trait. Lithics are also used to display individual traits, whether through varying raw materials or different places in the grave. The child has a limestone flake near the skull, whereas the two adult males with flints have flint or a red stone near their pelves. Unworked stones are deposited in the fill of three graves – the only child and two out of the four males. Finally, shells are placed on the arm of one of the adult males. In Group O, there is a strong sense of group identity in burial rites, alongside age-based, sex-based as well as many individual traits in the grave goods.

Group P (Fig. 48) This small Group is an all-male set, the only such Group in the excavated sample from Budakalász. The two Group traits are contracted inhumation in oval graves. Body position, grave orientation and grave depth form contrasts between each of the two pairs. Since there is only a single grave good deposition – two flint blades near the skull of the first burial – there is little contrast in this domain. Hence, the overall scale of contrasts is high, ranging from 50 to 70%, because of the variability in burial rites.

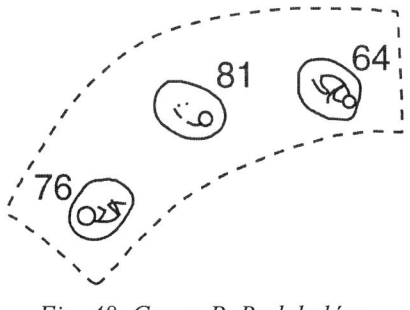

Fig. 48. Group P, Budakalász.

Group Q (Fig. 49 & Table 50) Group Q forms a strong contrast to Group P, both in terms of size, composition and inner complexity. It also forms a distinctive curved shape. With one double grave (a child and an adult male), the group contains more adult females than adult males and children. There are three putative Group traits, suggesting strong Group identity: contracted inhumation in seven out of the eight cases where there is information: oval graves in all except the rectangular double grave; and similar grave depth to within 10 cm (85-95 cm), except for the very shallow first grave at the E end. Body sidedness may be structured in a linear way, with right-sidedness found in the E part of the line, left-sidedness in the W part. Grave orientation provides the opportunity for most individual variation, with contrasts between every successive pair of burials.

The scale of contrasts is low, ranging from 10 to 45%, based upon 17 traits. The lowest score occurs in the double grave, contrasting a child and an adult male; the highest contrasts the preceding burial of an adult female and the same child.

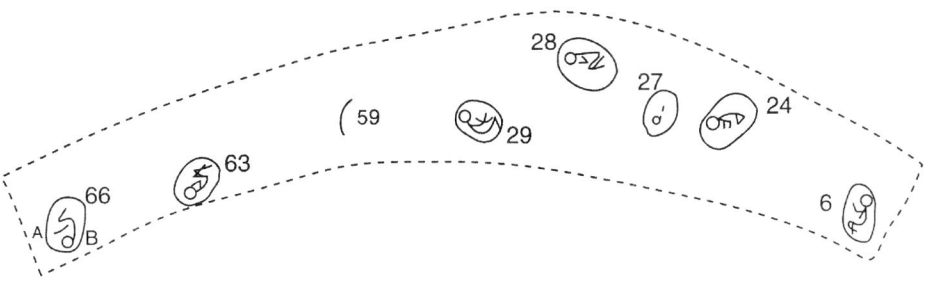

Fig. 49. Group Q, Budakalász.

Table 50. Grave good, costume and animal bone contrasts in Group Q, Budakalász.

Grave 66	Grave 63	Grave 59	Grave 29	Grave 28	Grave 27	Grave 24	Grave 6
necklace	necklace	-	copper applique	copper applique	-	shell beads	-
shell plaque	plaque						
	shell anklet						
-	-	bowl, jugs	-	dipper	-	-	body sherd
		dipper					
unworked stone	-	-	-	large stone	unworked stones	unworked stones	-
				red stone		jasper flake	
-	copper needle	-	-	-	-	-	-

155

The variability in grave goods can all be related to individual differentiation, with no sign of any categorical contrast. A complex set of variation in costume elements occurs, not least in the double grave, with shell plaques on the hips of the child and a shell bead and *Dentalium* necklace on the adult male. A similar necklace and a similar plaque (though placed on the scapula) adorn another adult female, who also has a shell bead anklet and a copper needle in her hand. A rare case is the attachment of shell beads to clothing on the back of an adult female. Similar shell appliques are found attached to the scapulae of both an adult male and an adult female. By comparison, relatively few ceramics are deposited in this Group. Body sherds are placed on the arm of an adult female, a dipper by the scapula of an adult male and a bowl, two jugs and a dipper in the fill of disturbed grave 59. Unworked stones are placed in the fill of three graves – the double grave, a child and an adult female. The same adult female has a jasper flake placed on her leg, while an adult male has a large unworked stone and a red stone placed on his pelvis. In summary, this Group exhibits much individual variation in grave good deposition, with a strong element of group identity in burial rites.

Group R (Fig. 50) The alignment of this small Group differentiates it from all other Groups – a straight line from NE to SW. Each age-sex category is represented, with more adult females than others. Several Group traits are recognisable: contracted inhumation on the left side, mostly in oval graves (three cases out of four) and with grave depth increasing in linear fashion. Grave orientation provides contrasts between each successive pair of graves. The scale of contrasts is moderate, ranging from 40 to 60% based upon 10 traits. The lower score distinguishes two pairs of graves – two adult females and the only child and an adult male, while the higher score contrasts an adult female and the child.

There are two cases of age-sex differentiation: body sherds are found in the fill of the adult female graves only, while, in the only case in the entire excavated sample, pointed stones are found in the side of the grave fill of the adult male – perhaps indicating a re-deposition of a grave marker.

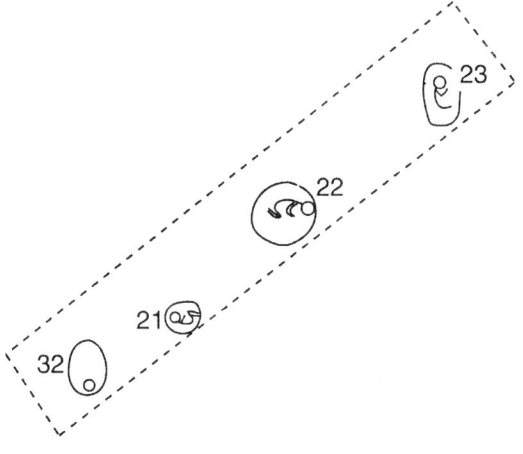

Fig. 50. Group R, Budakalász.

Likewise, an age-based trait concerns the unworked stones placed in the graves of all the adults but not the child. Individual differentiation is limited to one case of lithics (a hydroquartzite flake in one of the adult female graves) and to costume elements, where similar shell bead necklaces are found in the graves of the child and one of the adult females and a *Dentalium* and polished stone bead necklace adorns the other adult female. The variability in this Group is partially related to categorical traits of both age and age/sex, with some individual traits and a strong sense of group identity displayed in the burial rites.

5.4 Discussion

The excavated sample of the Budakalasz cemetery provides a clear pattern of the five different forms of variability, which have already been identified at the Basatanya cemetery (see chapter 4). Group identity is established almost entirely through the use of distinctive burial rites, of which five have been analysed. The three forms of categorical trait – age/sex, sex-based and age-based – are poorly represented at Budakalász, in comparison with individual traits, which are commonly encountered and lead to complex patterns of variability.

There is only one Group trait which relates to grave goods – the negative trait of the absence of pottery deposition in Group I. The commonest Group trait is actually so common that it stands for the identity of the whole community, rather than for that of any particular Group. This is the contracted inhumation rite, a defining characteristic in 15 out of the 18 grave lines. In the case of the exceptions, Groups C and M are small groups with several different burial modes, while Group I is defined by the only use in the cemetery of burning in the grave after deposition of bones. The next commonest Group trait, occurring in half of the grave lines, is the oval grave form. Since the distribution of this grave form is confined to the Northern and Western zones of the cemetery, it is likely that we can specify a sixth source of variability in the cemetery – viz. zonal identity, pertaining of most of the Groups in certain zones. Another candidate for zonal identity is the wave pattern found in grave depth, limited to lines mostly in the South East zone (four out of five cases) but also a few in the Northern zone and even one in the Western zone. However, this pattern occurs in only seven Groups and the distribution is perhaps insufficiently concentrated to form a valid zonal indicator.

The rare occurrence of other Group traits make them more valuable markers, especially in the case of single occurrences, such as the wave pattern of grave form found in Group A, the SSE-NNW grave orientation found in Group B, the right-sided body position of Group D the left-sided body position of Group O and the grave-base burning of Group I. Multiple occurrences of Group traits are also valuable, as with the linear change in grave depth in Groups H and R, and the narrow range of grave depths (less than 10 cm) found in Groups D, J and Q. The significance of Group, zonal and community traits is shown below (Table 51):

Table 51. Distribution of Group, Zonal and Community Traits
by Grave groups, Budakalász.

NO. OF TRAITS	0	1	2	3
GROUPS	M	C, L	E, F, K, N, P	A, B, D, G, H, I, J,O, Q, R

It is clear that over half of the Groups are characterised by the maximum number of traits referring to a larger entity – whether a Group, a Zone or the whole cemetery. Only the small Group M is not defined by a single Group trait and few Groups are characterised by only one such traits.

The paucity of categorical traits has been apparent through the analyses of the grave Groups. Only three Groups exhibit a total of four age/sex-based traits, all relating to grave goods – one using costume (adult female), one using ceramics (adult female) and two using lithics (both adult males). The situation is little different for sex-based traits, found as a total of six traits in five Groups. This total, however, includes Group M, with its two identifiable graves. Group C uses the grave form as a sex-related indicator but the others relate to grave goods. Most of the traits relate to costume, sometimes found with all adult males (Group C and O), sometimes found with all adult females (Groups I and M). Pottery is also variably associated with adult females in one group (A) and with males in another Group (C). Lithics are found only in a single Group (males in C). With the total of six age-related traits found in five Groups, three instances concern unworked stones found in grave fill – twice with all children's graves (Groups A and L), once in all adult male's graves (Group R). Other varied use of the same resources relates to costume, found in no adult male graves (Group L) and in no children's graves (Group O). Adult female graves in Group H are characterised by deposition of a bowl. The pattern is for varied use of the same, widely available material culture to convey differing messages in different contexts within the same cemetery.

The overall picture of the rarity of categorical traits as compared to group traits is demonstrated below (Table 52):

158

Table 52. Distribution of Categorical Traits by Grave groups, Budakalász.

NO. OF TRAITS	0	1	2
GROUPS	B, D, E, J, K, N, P, Q	C, F, G, H, I, L, M	A, O, R

This distribution contrasts very strongly with the Group trait distributions in Table 51, indicating that the negotiation of age/sex categories took a very different form from the ways in which small-group identities developed.

The principal source of variability in the Budakalász grave Groups is the individual traits which are so significant in almost all of the Groups. Table 53 provides a summary of the overall distribution of individual traits by domain:

Table 53. Distribution of Individual Traits by Grave groups, Budakalász.

NO. OF TRAITS	0	1	2	3	4
BURIAL RITE	M	B, D, H, I, J, K, O, Q, R	C, F, G, L	E, P	A, N
COSTUME	C, L, M, N, P	A, B, J, K, R	E, F, G, O		D, H, I, Q
POTTERY	C, I, M, P, R	A, D, F, G, K, O	J, L	B, N, Q	E, H
OTHER	B, L, M, N	C, D, G, P, R	A, I	E, F, H	J, K, O, Q

This indicates that there is no overall zonal difference between the use of individual traits in respect of the specified domains but rather that it is the individual grave Groups where decisions are made about most differentiation between successive pair of burials. A summary of the distribution of individual traits by grave Group follows (Table 54):

Table 54. Summary of Individual Traits by Grave groups, Budakalász.

NO. OF TRAITS	0	1	2	3	4
GROUPS	M		C, L, N, P	B, I, R	A, D, E, F, G, H, J, K, O, Q

The distribution in this Table resembles the pattern of Group traits (Table 51), with over half of the Groups exhibiting the maximum number of classes of individual traits – a clear indication of the importance of individual sources of variability in the cemetery as a whole.

Each of the many sources of mortuary variability at Budakalász is available to each of the small kinship groups who had primary responsibility for the decisions on how to bury their newly-dead and with which grave goods.

It appears that the vast majority of Groups selected a range of Group traits from the burial rite and an even wider and more complex suite of grave goods for individual traits.

The quantitative analysis of the scale of contrasts used in each Group is summarised below (*Fig. 51*). There is much variation in the scale of contrasts between Groups, which does not appear to be correlated to the number of traits or to the size of the Group. Zonal patterning also shows up in this graph, too: the lowest range of difference occurs in the Northern zone, with little difference between the Western and South Eastern zones. But there is much overlap in all three zones, suggesting that the decisions about how to express contrasts and distinctions between successive pairs of graves remained at the local level, with little if any reference to higher-level decision-making at the zonal level.

Another way to examine such a claim is the investigation of how global rules pertaining to the whole cemetery are put into practice in the local context of individual grave Groups. The global rules of preference for contracted inhumation in oval graves were amply corroborated, but the role of rectangular graves was much less important in the vast majority of Groups than seemed likely from the perspective of the whole cemetery. Interestingly, none of the other global rules is contradicted by the micro-tradition analysis. Instead, different uses were made of the common material culture to emphasise varying forms of identity, from the individual to the zonal.

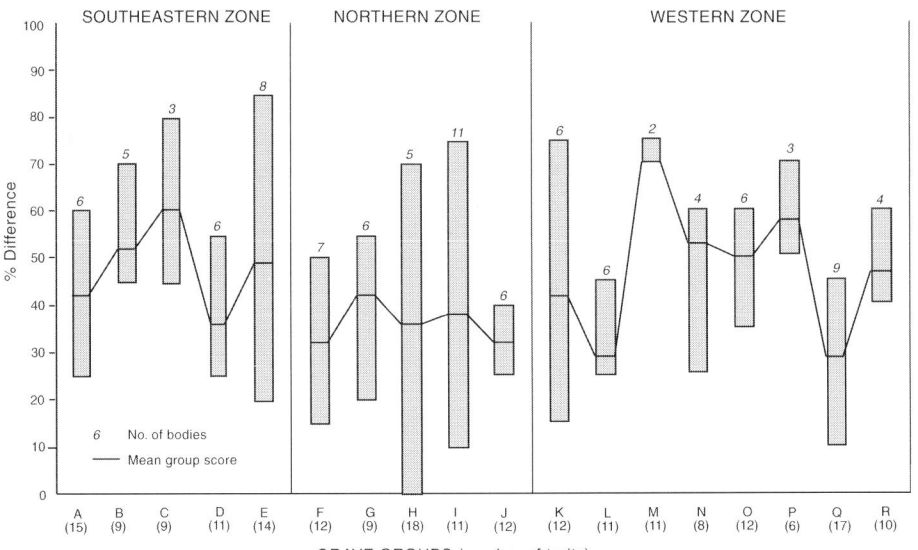

Fig. 51. Range of contrasts in mortuary practices by Group, Budakalász.

5.5 Summary

The Budakalász cemetery is a good example of a cemetery whose spatial order, internal variability and external contacts are guided by fuzzy rules rather than polar opposites in mortuary wealth and diversity. Whittle's (1996: 124) characterisation of Budakalász as "unspectacularly furnished with grave goods" is correct but misses the point. What is represented here is not the extremes in mortuary wealth of a Varna I cemetery but the overlapping categorisations of a community where combinations of raw materials and of grave goods classes together with variability in location of grave goods provides kaleidoscopic potential for fine-tuned and diverse statements about the living and the dead. The variability in costume and grave accoutrements are not being used to deploy the wealth of the community, nor is the message of the age-grades or the sex roles locked in dispute. Instead, women, men and children partake of similar but subtly varied identities, of the kind which signals a society where major change is not an issue in the same way that it was negotiated at Basatanya. The identities of these individuals are defined in time and space through the burial rites and material culture depositions which elaborate upon the ordering within the linear groupings which comprise their most intimate kinship circles. In none of the mortuary assemblages examined in this research has there been found such strong emphasis on traits exhibiting individual identity as at the Budakalász cemetery. This statement about the importance of individual members of the kinship grouping is not vitiated by the paucity of material culture used to stress categorical divisions – age/sex, age-based or sex-based identities. But the importance of the wider kinship network is also signalled by the widespread use of burial rites to emphasise corporate group identity.

6. Conclusions

6.1 Theory and method

I started this book with three imaginary scenes – one from each of the prehistoric epochs which has supplied the raw materials for the analysis of micro-traditions. Those scenes were included not so much to provide a verifiable story of past events, though each of them is actually rather believable, but to emphasise one of the main themes of the book. That theme is that people were social actors who created not just the major and significant highlights of what became cultural memory – what I have called 'timemarks' – but also the longer-term habitus through the repetition of those customary, small-scale, day-to-day events which were not spoken about in feasts or ceremonies. These people worked to produce themselves through their everyday artifacts and structures – the material remains of which are the subject matter of prehistoric archaeology.

I have tried to develop the theme of social structuring through a methodology which avoids the central weakness of the Giddensian world-view – the reification of the structure-agency duality. The key concept here is self-categorisation, which is a type of agency at the same time as being a structuring device. In the approach termed dynamic nominalism, categories of people come into being at the same time as the kinds of people who fit those categories. In this way, the continuous construction of society is itself a strategy, open to challenges from different people and full of the tensions of any debate about power. Hence, agency and structure come together in the formation of identities – the practices of self-description through categorisation. Cultural identities are therefore seen as public acts of mediation between the self and others; self-definition is seen as a selection from one's own history and origins – a narrative about inclusions and exclusions.

Danny Miller's theory of culture provides a way of integrating the dynamic nominalist approach to the world of material culture. In this, people achieve differentiation through objectification – the production of objects and structures – but cannot prevent alienation from these things without the complementary process of sublation or the re-incorporation of the material world into cultural memory and practices. Hence, the subject of culture – people themselves – are reflexively constituted, as in the self-description of people as categories. For Miller, categorisation enables objects to integrate individuals within a normative social order; objects tell insider stories about the group's categorisation processes. I have made similar comments about the importance of places relative to people in a mutually constitutive creation of identity.

It is a curious fact that, although enormously popular in current archaeology, mortuary studies has been heavily under-theorised with respect to the structure – agency debate. Curious, because the conditions for such a project are generally present. Put at its simplest, each new burial, viewed as action, presents the challenge to the living of how to integrate the newly-dead into the structured society of the ancestors already present in the settlement or cemetery (structure). All previous burials have created a cumulative structure, in terms of the traditions selected and given material form. The choice for the living at the time of a new death is to what extent to confront and challenge the existing structure or maintain the status quo by adopting the standard burial rites of the community. Framed in Millerian terms, the final journey of the newly-dead is a representation of what the living wish to say about their own views of the world and about that person and their community – an externalisation prior to the final sublation, both for the living and the newly-dead, at the burial site – a *locus classicus* of consumption.

In traditional mortuary analysis, it has been customary to select one of two units of analysis: the individual grave or the total buried population. An exception is the selection of two or more main phases of the total buried population, as in Tiszapolgár-Basatanya. The problem with the selection of the individual grave is an excess of empirical detail, leading to an indigestible presentation of undigested facts. Equally, the difficulty with using the whole population, or chronological sub-divisions thereof, is that the spatial and temporal framework of the **succession** of burial rites is abolished, rendering the analysis ahistorical.

I believe that this problem has been tacitly accepted because it covered up a more basic problem – the inability of archaeologists to define internal sequences of burials in prehistoric cemeteries. The method which I term 'micro-tradition analysis' is an attempt to use the best evidence we have for working with the internal succession of burials – in settlements and cemeteries where graves are structured in lines. Burial lines can form the standard spatial pattern for mortuary practices in some regions (e.g. Early Medieval Germany) and can occur sporadically in many other place/times, from Japanese Jomon to Roman Britain via Copper Age Italy. But, as far as I am aware, there has been no analysis which treats the grave line as the primary unit of analysis, while working in conjunction with studies of the total buried population and the individual graves.

A set of criteria is derived for the optimal operation of a micro-tradition analysis, which involves the minimal number of graves sufficient to form a line, the integrity of the grave line, the unidirectionality of the burial sequence, the starting point for grave lines and the predominance of graves in lines over all other forms of spatial ordering. The evidence required to be able to demonstrate the feasibility of a micro-tradition analysis at a specific site is often readily available for careful evaluation of the site potential. If the feasibility of a site can be demonstrated, the potential gains can be high, since the selection of the grave line as a unit of analysis permits a detailed examination of the sources of variability for each grave. Five sources – relating to different kinds of identities – have been established: community identity (relating to the global rules applied to mortuary rites), zonal identity (in large mortuary domains with spatial clustering of grave lines), corporate group identity (where traits are found in all or most of the graves in a specific line and not in other lines), age-sex identity, age-only identity, sex-only identity and individual identity (to account for differences between two or more graves of the same or similar age-sex characteristics).

On this basis, I attempted a micro-tradition analysis of the three most suitable published sites in later Hungarian prehistory – the Late Neolithic intramural burials at Kisköre-Damm, the Earlier Copper Age cemetery of Tiszapolgár-Basatanya and the Late Copper Age cemetery of Budakalász-Luppa csárda.

6.2 Results

The Kisköre-Damm burials are structured around two spatial principles – the prior existence of a Middle Neolithic grave line, or 'ancestral line' and the grouping of the majority of Late Neolithic graves into burial lines, most of which are related to Late Neolithic Tisza houses. Earlier categorical analysis of the Kisköre burials had led to the definition of eight global rules of mortuary practice, as related to the analysis of the total buried population. The global rules emphasised the importance of fluid, cross-cutting categorisation to the Kisköre community rather than the sharper, binary oppositions which could have been utilised. However, the micro-tradition analysis suggests that, although all the major dimensions of mortuary variability could be related to Group identities, very few related to the global rules. Two cases were found of burial lines whose defining trait ran counter to the global rule – both related to the degree of costume elaboration. If, as is spatially and socially highly probable, most of the grave lines at Kisköre are integrally related to households, this result suggests that Sahlins' (1972) famous claim that the wider community was the household writ large may have been wrong. Here at Kisköre, the local is certainly more than the global writ small!

There are relatively few examples of identities which can be referred to age and sex identity, age-only identity or sex-only identity. In most cases, the costume elements and grave goods are widely cross-referenced to other graves in other lines. But several grave lines show signs of a strong Group identity because of the importance of directional traits, which, by definition, illustrate the co-emergence of group structure and personal identity.

In the spatially distinct cemetery of Tiszapolgár-Basatanya, a far larger sample of burials and a much more complex spatial structure makes the analysis of micro-traditions more difficult than at Kiskőre. These complexities are reflected in the large body of global rules – 17 in all – which structures burials in the two Periods of Basatanya (Period I – Early Copper Age Tiszapolgár group; Period II – Middle Copper Age Bodrogkeresztúr group). A continuing problem concerns the total duration of burial activity at Basatanya – whether 200 or 800 years. No advance on this problem was made in this analysis.

A strong case was made out that an antecedent Late Neolithic burial provided the point of reference for the start of Period I burials. Eight grave lines were identified in Period I, each with a varied age-sex composition and individuals suffering from skeletal pathologies. A wider artifactual range than at Kiskőre (from 3 to 9 categories, depending upon grave line) was available to draw upon for material differentiation.

Five forms of identity could be recognised in Period I: no zonal traits could be distinguished. There was widespread emulation of the global rules in most grave lines, especially in burial orientation. The paucity of unique corporate group traits in Period I suggests the significance of links between different grave lines. Categorical traits showing age and/or sex identities were differently distributed at Basatanya fom those at Kiskőre. Age-related traits were commonest, then sex-related traits, with very few apparent age-and-sex-related traits. Individual differences were strongly represented, for children, adult males and adult females, but with different emphases on the various age-sex categories in different grave lines.

In Period II at Basatanya, 12 grave lines were created but a larger proportion of burials than in Period I was located elsewhere. Fewer grave lines than before contained each of the age-sex categories and fewer individuals with pathological injuries were buried in Period II.

Of the same five forms of identity traits found in Period I, two corporate group traits were shared between several grave lines – standard body position and orientation. Far more unique group traits are found than in Period I, suggesting the wider impact of different social networks upon local cemetery practices. The incidence of categorical traits differs strongly from that of Period I, with far more age-sex traits, fewer sex-related traits and as many age-related traits. The paucity of sex-related identity traits, unexpectedly, ran counter to Basatanya analyses since Bognár-Kutzián. A wide range of material culture was used to betoken individual identities in Period II – especially adult female identities. These traits are found in over half the grave lines, ranging from very intensive in some lines to very restricted in others.

In both Periods at Basatanya, there is surprisingly little variation from global rules in the grave lines; more deviations are found in the later Period. However, some grave lines in each Period exhibit major differentiation from standard global rites. Here, it is important to note that the same item of material culture is used to signal different messages in various grave lines. This trend is paralleled in the Copper Age cemeteries of North East Bulgaria (Chapman 1996).

In summary, the general trend in Period I at Basatanya is the emergence of a stable pattern of community burial, with global rules in evidence in all of the different grave lines. The rules often highlight the oppositions between age-sex categories, although cross-cutting individual traits are also common. This pattern changes in Period II, with greater differences between grave lines and hence a wider incidence of deviation from global rules. The implications are that a wider range of social networks has been established, involving people from a wider area, who use a wider range of material culture, and whose kin relations have varying modes of self-categorisation and self-identification.

The partial sample from the Budakalász cemetery has no published mortuary antecedents – a severe limitation on understanding the directionality of burial lines. Nonetheless, limited stratigraphic data from inter-cutting graves allow the inference of a SE-NW direction for the cemetery as a whole. The published graves cluster into three main zones – the West with 8 grave lines, the North with 5 lines and the South East with 5 lines.

The suite of nine global rules derived from a categorical analysis of Budakalász indicates a far looser structure for burial rites than that created at Basatanya. The fluid, cross-cutting pattern resembles far more closely the global rules of Kisköre-Damm. Two major differences from Basatanya are the much higher incidence of children and multiple graves at Budakalász. This leads to a greater variety in age-sex composition of grave lines than at either Kisköre or Basatanya.

The distribution of identity traits at Budakalász is strikingly different from those of the other two sites. Only one community-wide trait is found – the prevalence of standard contracted inhumations in 15/18 grave lines. Two zonal traits can be identified with a fair degree of confidence – the oval grave form in the N and W Zones and the wave pattern in grave depths, predominantly (but not exclusively) found in the SE Zone. Group traits are rarely unique as far as either grave goods or mortuary rites are concerned. Age-and-sex-related identity traits are very rare, sex-related traits are quite rare and age-related traits only a little more frequent. The key domain for identities is the individual trait, which is important in by far the majority of grave lines. The scale of Group contrasts is generally high – higher than at Basatanya – possibly related to Zonal patterning (lower contrasts in the W zone, higher in the N and SE).

The pattern of fuzzy rules, showing overlapping variability rather than polar opposites, is found in the micro-tradition analysis as much as in the earlier categorical analysis. The extent of variability at Budakalász is such that the main loci for the expression of change are local grave lines, with a strong emphasis on individual identities.

6.3 Wider implications and future research directions

The specific results of the categorical and micro-tradition analyses of three mortuary sites from later Hungarian prehistory are surely not the stuff of grand narrative – the re-writing of the normative structure of the Neolithic and the Copper Age of the Carpathian Basin in an innovative way. It should be admitted that this was never the aim of the study. No-one is currently in a position to assess the probability that the three sites analysed here are in any way comparable to other coeval mortuary groupings, for the simple reason that the requisite analyses have not yet been tried. It is therefore a huge assumption that Kisköre, Basatanya and Budakalász are typical of other mortuary sites of their epoch. But, until we have a range of comparable analyses, it is an assumption that is tempting to make.

Instead, I hope that a suite of more limited aims has been satisfied – the theorising of a new approach to mortuary data, the development of an embryonic methodology for the recognition of sites eligible for the new approach, and a trial run using three data sets of differing complexities and suitabilities. Are there any lessons we can draw from the manifest differences between Kisköre, Basatanya and Budakalász?

Undoubtedly the biggest difference between the three sites lies in the predominant form of categorisation. Sandwiched between two time/places where fluid, cross-cutting categorisation was developed with material means and through a complex set of mortuary rites, Basatanya appeared to rely more on oppositional categorisation for the differentiation of both age-based and sex-based identities. While it can be argued that major changes took place in each of the three periods in question (for a summary, see Chapman 1994), there is a strong case that the introduction of copper and gold metallurgy against a background of major settlement change and the introduction of novel secondary products made the Earlier Copper Age one period of major change in later

Hungarian prehistory. The traditional, categorical analysis of the Basatanya cemetery reinforced the link between binary opposites in categorisation and major socio-economic change. But the micro-tradition analysis raises questions about this link by highlighting the full range of practices by which identities were constituted at Tiszapolgár. If the link between separate grave lines and different co-residence-based corporate groups can be substantiated, the roots of cultural change may be more closely related to variability between households or farmsteads rather than to diachronic change across the *longue durée*.

The other more general observation is the ways in which the same widely available items of material culture were consistently used to signal different messages in different grave lines. This was documented at each of the three sites and should therefore be considered as a well-established pattern. This finding is reminiscent of the way in which the three different North East Bulgarian Late Copper Age communities whose cemeteries were studied using categorical analysis were found to use the same objects in quite different ways in each community (Chapman 1996). A strong pattern of multiple uses was found for ornaments, tools and pottery, such that only one single artifact type out of over 50 was used to identify the same age-sex categories in all three cemeteries (the carinated bowl). Additionally, a comparison of artifacts deposited on the tell and in the adjacent cemetery at Goljamo Delchevo indicated that differences existed at the sub-type level for every major artifact type. Here, the inclusions and exclusions characteristic of the narrative of identity-formation suggest two rather different modes of deposition, probably by two different corporate groups. These results emphasise the need to make further studies on the different ways in which individuals and corporate groups draw upon the same range of available material culture in pursuit of self-categorisation and self-identity.

Further research is clearly necessary to test the wider applicability of micro-tradition analysis. For instance, it may be that the analysis is not limited to mortuary zones with pronounced grave lines. An important point is that the availability of a large burial sample is not necessary for a successful micro-tradition analysis, which is suitable for small cemeteries as well as large. Another research area where further work is urgently required is the establishment of principles for the construction of sound internal chronologies for the intra-mural mortuary zone and the cemetery. It seems unlikely that even the availability of high-resolution 14-C dates for all the lines of Basatanya would resolve the dating problems raised in this study. But dating projects would be wise to take account of grave line distributions in the selection of human bone for future direct dating.

Finally, it is imperative to gain a more detailed and nuanced account of the physical anthropology of these classic Hungarian prehistoric mortuary populations. The largest current constraint upon both categorical and micro-tradition analyses is the very general age-sex categories utilised by earlier physical anthropologists. A re-analysis of these collections would yield huge dividends for both anthropological and cultural research.

References

ARSENAULT, D. 1991
The representation of women in Moche iconography. In: Walde, D., Willows, N. D. (eds.), *The archaeology of gender*. Calgary, University of Calgary Archaeological Association, 313-326.

BAILEY, D. W. 1996
The life and times of House 59,Tell Ovcharovo. In: Darvill, T., Thomas J. (eds.), *Neolithic houses of NW Europe and beyond*. Oxford, Oxbow, 143-156.

BAILEY, D. W. (ed.) 1998
The archaeology of prestige and wealth. International Series 730. Oxford, BAR, 106-130.

BÁNFFY, E. 1990/91
Cult and archaeological context in Central and South East Europe in the Neolithic and Chalcolithic. *Antaeus* 19-20, 183-250.

BÁNFFY, E. 1995
South-West Transdanubia as a mediating area. On the cultural history of the Early and Middle Chalcolithic. *Antaeus* 22, 157-196.

BÁNFFY, E. 1997
Cult objects of the Neolithic Lengyel culture. Connections and interpretations. Budapest, Archaeolingua, Series Minor 7.

BANNER, J. 1937
Die Ethnologie der Körös-Kultur. *Dolgozatok* 13, 32-49.

BANNER, J. 1956
Die Peceler Kultur. Archaeologia Hungarica 35. Budapest, Akadémiai Kiadó.

BARRETT, J. 1988
Fields of discourse. Reconstituting a social archaeology. *Critique of Anthropology* 7/3, 5-16.

BARRETT, J. 1990
The monumentality of death. *World Archaeology* 22/2, 179-189.

BARRETT, J. 1994
Fragments from antiquity. An archaeology of social life in Britain, 2900 - 1200 BC. Oxford, Blackwell.

BARTOSIEWICZ, L. 1984
An attempted distinction between the parts of the Neolithic site at Csabdi-Telizöldes. *Acta Archaeologica Hungarica* 36, 43-52.

BARTOSIEWICZ, L. 1995
Archeozoological studies from the Hahot Basin, SW Hungary. *Antaeus* 22, 307-380.

BARTOSIEWICZ, L. 1998
Sándor Bökönyi: Portrait with a scientific background. In: Anreiter, P., Bartosiewicz, L., Jerem, E., Meid, W. (eds.), *Man and the Animal World.* Budapest, Archaeolingua, 3-13.

BEAUDRY, M. C. – COOK, L. J. – MROZOWSKI, S. A. 1991
Artifacts and active voices: material culture as social discourse. In: McGuire, R. H., Paynter, R. (eds.), *The archaeology of inequality.* Oxford, Blackwell, 150-191.

BECK, L. A. (ed.) 1995
Regional approaches to mortuary analysis. New York, Plenum Press.

BENKŐ, L. – HORVÁTH, F. – HORVATINČIĆ, N. – OBELIĆ, B. 1987
Radiocarbon and thermoluminescence dating of prehistoric sites in Hungary and Yugoslavia. *Radiocarbon* 31, 992-1002.

BINFORD, L. 1971
Mortuary practices: their study and potential. In: Brown, J. A. (ed.), *Approaches to the social dimensions of mortuary practices.* Society for American Archaeology Memoirs No. 25. Washington, D.C., SAA, 6-29.

BÍRÓ, K. 1986
International conference on prehistoric flint mining and lithic raw material identification in the Carpathian Basin. 2 vols. Budapest, Magyar Nemzeti Múzeum.

BÍRÓ, K. 1987
Chipped stone industry of the Linearband Pottery Culture in Hungary. In: Kozłowski, J. K., Kozłowski, S. K. (eds.), *Chipped Stone Industries of the Early Forming Cultures in Europe.* Warszawa, Wydawnictwa Universytetu Warszawskiego, 131-167.

BÍRÓ, K. 1988
Distribution of lithic raw materials on prehistoric sites. An interim report. *Acta Archaeologica Hungarica* 40, 251-274.

BÍRÓ, K. 1992
Adatok a korai baltakészítés technológiájához. *Acta Musei Papensis* 3-4, 33-80.

BÍRÓ, K. 1998
Lithic materials and the circulation of raw materials in the Great Hungarian Plain during the Late Neolithic period. Budapest, Hungarian National Museum.

BLAKE, E. 1999
Identity mapping in the Sardinian Bronze Age. *European Journal of Archaeology* 2/1, 35-55.

BOGNÁR-KUTZIÁN, I. 1963
The Copper Age cemetery of Tiszapolgár-Basatanya. *Archaeologia Hungarica* 42. Budapest, Akadémiai Kiadó.

BOGNÁR-KUTZIÁN, I. 1972
The Early Copper Age Tiszapolgár culture in the Carpathian Basin. Budapest, Akadémiai Kiadó.

BÖKÖNYI, S. (ed.) 1992
Cultural and landscape changes in South East Hungary I. Reports on the Gyomaendrőd Project. Budapest, Archaeolingua.

BÖKÖNYI, S. 1992a
Introduction. In: Bökönyi, S. (ed.), *Cultural and landscape changes in South East Hungary I. Reports on the Gyomaendrőd Project.* Budapest, Archaeolingua.

BÖKÖNYI, S. 1992b
The Early Neolithic vertebrate fauna of Endrőd 119. In: Bökönyi, S. (ed.), *Cultural and landscape changes in South East Hungary I. Reports on the Gyomaendrőd Project.* Budapest, Archaeolingua, 195-300.

BÖKÖNYI, S. 1993
Recent developments in Hungarian archaeology. *Antiquity* 67, 142-145.

BOURDIEU, P. 1977
Outline of a theory of practice. Cambridge, Cambridge University Press.

BUKO, A. 1998
Pottery, potsherds and the archaeologist: an approach to pottery analysis. In: Tabaczyński, S. (ed.), *Theory and practice of archaeological research.* vol III. Warszawa, Institute of Archaeology and Ethnology, 381-408.

CHAPMAN, J. C. 1977
The Balkans in the Fifth and Fourth Millennia BC. Unpublished PhD thesis, University of London.

CHAPMAN, J. 1981
The Vinča culture of south east Europe. Studies in chronology, economy and society. 2 vols., International Series 119. Oxford, BAR.

CHAPMAN, J. C. 1982
The secondary products revolution and the limitations of the Neolithic. *Bulletin of the University of London Institute of Archaeology* 11, 119-141.

CHAPMAN, J. C. 1983

Meaning and illusion in the study of burial in Balkan prehistory. In: Poulter, A. (ed.), *Ancient Bulgaria.* Volume 1. Nottingham, University of Nottingham Press, 1-45.

CHAPMAN, J. 1989

The Early Balkan village. *Varia Archaeologia Hungarica* 2, 33-53.

CHAPMAN, J. 1991

The creation of social arenas in the Neolithic and Copper Age of South East Europe: the case of Varna. In: Garwood, P., Jennings, P., Skeates, R., Toms, J. (eds.), *Sacred and Profane.* Oxford Committee for Archaeology Monograph No. 32. Oxford, Oxbow, 152-171.

CHAPMAN, J. C. 1993

Social power in the Iron Gates Mesolithic. In: Chapman, J. C., Dolukhanov, P. M. (eds.), *Cultural transformations and interactions in Eastern Europe.* Worldwide Archaeology Series 5. Aldershot, Avebury, 61-106.

CHAPMAN, J. C. 1994

The living, the dead, and the ancestors: time, life cycles and the mortuary domain in later European prehistory. In: Davies, J. (ed.), *Ritual and remembrance. Responses to death in human societies.* Sheffield, Sheffield Academic Press, 40-85.

CHAPMAN, J. 1995

Social power in the early farming communities of Eastern Hungary – perspectives from the Upper Tisza region. *A Jósa András Múzeum Évkönyve* 36, 79-99.

CHAPMAN, J. 1996

Enchainment, commodification and gender in the Balkan Neolithic and Copper Age. *Journal of European Archaeology* 4, 203-242.

CHAPMAN, J. 1997

Changing gender relations in the later prehistory of Eastern Hungary. In: Moore, J., Scott, E. (eds.), *Invisible people and processes. Writing women and children into European archaeology.* Leicester, Leicester University Press, 131-149.

CHAPMAN, J. 1997a

Places as timemarks – the social construction of landscapes in Eastern Hungary. In: Chapman J., Dolukhanov, P. (eds.), *Landscapes in Flux.* Colloquenda Pontica 3. Oxford, Oxbow Books, 137-162.

CHAPMAN, J. 1998

Objects and places: their value in the past. In: Bailey, D. W. (ed.), *The archaeology of prestige and wealth.* International Series 730. Oxford, BAR, 106-130.

CHAPMAN, J. C. 2000

Fragmentation in archaeology: people, places and broken objects in the prehistory of South Eastern Europe. London, Routledge.

CHAPMAN, J. 2000a

Tension at funerals: social practices and the subversion of community structure in later Hungarian prehistory. In: Dobres, M.-A., Robb, J. (eds.), *Agency in archaeology.* London, Routledge, 169-195.

CHAPMAN, J. – LASZLOVSZKY, J. 1992

The Upper Tisza Project 1991: report on the first season. In: *Archaeological Reports 1991* (Durham, Newcastle upon Tyne). Durham, University of Durham, 10-13.

CHAPMAN, J. – LASZLOVSZKY, J. 1993

The Upper Tisza Project: the September 1992 season. In: *Archaeological Reports 1992* (Durham, Newcastle upon Tyne). Durham, University of Durham, 13-19.

CHAPMAN, J. – LASZLOVSZKY, J. 1994

The Upper Tisza Project - September 1993 season. In: *Archaeological Reports 1993* (Durham, Newcastle upon Tyne). Durham, University of Durham, 1-7.

CHAPMAN, J. – LASZLOVSZKY, J. 1995

A Neolithic flood in eastern Hungary: the Upper Tisza Project 1994. In: *Archaeological Reports 1994* (Durham, Newcastle upon Tyne). Durham, University of Durham, 8-17.

CHAPMAN, J.– POLLARD, C. J. – PASSMORE, D. G. – DAVIS, B. A. S. 1997

Sites and palaeo-channels in the Polgár lowlands, North East Hungary: the Upper Tisza Project 1996 field season. In: *Archaeological Reports 1996* (Durham, Newcastle upon Tyne). Durham, University of Durham, 12-21.

CHAPMAN, J. – VICZE, M. 1996

Archaeological Reports 1995 (Durham, Newcastle upon Tyne). Durham, University of Durham, 34-40.

CHILDE, V. G. 1929

The Danube in prehistory. Oxford, Clarendon.

CHILDE, V. G. 1939

The dawn of European civilisation. 3rd (ed.), London, Kegan Paul.

CHOYKE, A. M. 1981

Régészeti lelőhelyek módszeres felszíni vizsgálata (Systematic archaeological survey). *Archaeológiai Értesítő* 108/1, 95-99.

CONNELL, R. W. 1987
Gender and power. Society, the person and sexual politics. Cambridge, Polity Press.

COX, M. 1996
Life and death in Spitalfields, 1700-1850. York, Council for British Archaeology.

CSALOG, J. 1959
Die anthropomorphen Gefässe und Idolplastiken von Szegvár-Tűzköves. *Acta Archaeologica Hungarica* 11, 7-38.

DOBRES, M.-A. 1995
Gender and prehistoric technology: on the social agency of technical strategies. *World Archaeology* 27/1, 25-49.

DOMBAY, J. 1960
Die Siedlung und das Gräberfeld in Zengővárkony. Budapest, Ungarische Akademie der Wissenschaften.

FIGLER, A. – BARTOSIEWICZ, L. – FÜLEKY, GY. – HERTELENDI, E. 1997
Copper Age settlement and the Danube water system: a case study from North-Western Hungary. In: Chapman, J., Dolukhanov, P. (eds.), *Landscapes in flux.* Colloquia Pontica 3. Oxford, Oxbow Books, 209-230.

FORENBAHER, S. 1993
Radiocarbon dates and absolute chronology of the central European Early Bronze Age. *Antiquity* 67, 235-256.

FOUCAULT, M. 1973
Madness and civilisation: a history of insanity in the Age of Reason. New York, Random House.

FOUCAULT, M. 1979
Discipline and punish: the birth of the prison. New York, Vintage Books.

FOUCAULT, M. 1984
Interviews. In: Rabinow, P. (ed.), *The Foucault reader.* London, Penguin.

FRIEDMAN, J. – ROWLANDS, M. J. 1977
The evolution of social systems. London, Duckworth.

GERO, J. – CONKEY, M. (eds.) 1991
Engendering archaeology. Oxford, Blackwell.

GIDDENS, A. 1987
Social theory and modern sociology. Cambridge, Polity Press.

GIDDENS, A. 1991
Modernity and self-identity: self and society in the Late Modern Age. Cambridge, Polity Press.

GILLINGS, M. 1997
Spatial organisation in the Tisza flood plain: landscape dynamics and GIS. In: Chapman, J. C., Dolukhanov, P. M. (eds.), *Landscapes in flux.* Colloquenda Pontica 3. Oxford, Oxbow Books, 163-180.

GILLINGS, M. 1998
Embracing uncertainty and challenging dualism in the GIS-based study of a palaeo-flood plain. *European Journal of Archaeology* 1/1, 117-144.

GILLINGS, M. – GOODRICK, G. 1996
Sensuous and reflexive GIS: exploring visualisation and VRML. *Internet Archaeology* 1, (http://intarch.ac.uk/journal/issue1/).

GIMBUTAS, M.1977
Gold treasure at Varna. *Archeology* 30, 44-51.

GOODMAN, R. B. 1995
Introduction. In: Goodman, R. B. (ed.), *Pragmatism: a contemporary reader.* London, Routledge, 1-20.

GYULAI, F. 1995
The plant and food remains from the Copper Age settlement at Zalaszetbalázs-Szőlőhegyi mező. *Antaeus* 22, 145-156.

HACKING, I. 1995
Three parables. In: Goodman, R. B. (ed.), *Pragmatism: a contemporary reader.* London, Routledge, 237-249.

HARTYÁNYI, B. – NOVÁKI, GY. – PATAY, A. 1968
Növényi mag- és termésleletek Magyarországon az újkőkortól a XVIII. századig. *A Magyar Mezőgazdasági Múzeum Közleményei* 1967-78, 5-84.

HARTYÁNYI, B. – SZ. MÁTHÉ, M. 1979
Pflanzliche Uberreste einer Wohnsiedlung aus dem Neolithikum im Karpaten-Becken. In: Körber-Grohne, U. (ed.), *Festschrift für Maria Hopf zum 65. Geburtstag am 14. September 1979.* Köln, Rheinland-Verlag GmbH, 97-114.

HAWKES, C. 1954
Archaeological method and theory: some suggestions from the Old World. *American Anthropologist* 56, 155-168.

HELMS, M. W. 1988
Ulysses' sail. An ethnographic odyssey of power, knowledge and geographical distance. Princeton, Princeton University Press.

HERTELENDI, E. – HORVÁTH, F. 1992
Radiocarbon chronology of late Neolithic settlements in the Tisza-Maros region, Hungary. *Radiocarbon* 34/3, 859-866.

HERTELENDI, E. – SVINGOR, É. – RACZKY, P. – HORVÁTH, F. – FUTÓ, I. – BARTOSIEWICZ, L. – MOLNÁR, M. 1998
Radiocarbon chronology of the Neolithic and time span of tell settlements in eastern Hungary based on calibrated radiocarbon dates. In: Költő, L., Bartosiewicz, L. (eds.), *Archaeometrical research in Hungary II.* Budapest – Kaposvár – Veszprém, 61-70.

HODDER, I. 1984
Burials, houses, women and men in the European Neolithic. In: Miller, D., Tilley, C. (eds.), *Ideology, power and prehistory.* Cambridge, Cambridge University Press, 51-68.

HODDER, I. 1990
The domestication of Europe. Oxford, Blackwell.

HORVÁTH, F. 1991
Vinča culture and its connections with the southeast Hungarian Neolithic: a comparison of traditional and 14C chronology. *Banatica* 11, 259-273.

HORVÁTH, F. – HERTELENDI, E. 1994
Contribution to the 14C based chronology of the Early and Middle Neolithic Tisza region. *A Jósa András Múzeum Évkönyve* 36, 111-133.

IVANOV, I. 1991
Der Bestattungsritus in der chalkolitischen Nekropole von Varna (mit einem Katalog der wichtigsten Gräber). In: J. Lichardus, J. (ed.), *Die Kupferzeit als historische Epoche.* Saarbrücker Beiträge zur Altertumskunde 55, Saarbrücken, Saarland Museum, 125-150.

JAMES, A. – JENKS, C. – PROUT, A. 1998
Theorising childhood. Cambridge, Polity Press.

JANKOVITCH, B. D. – MAKKAY, J. – SZŐKE, B. M. (eds.) 1989
Békés megye Régészeti Topográfiája. *Magyarország Régészeti Topográfiája* 8. Budapest, Akadémiai Kiadó.

JENKINS, R. 1997
Rethinking ethnicity: arguments and explorations. London, Routledge.

JOHNSON, M. 1989
Conceptions of agency in archaeological interpretation. *Journal of Anthropological Archaeology* 8, 189-211.

JONES, S. 1997
The archaeology of ethnicity: constructing identities in the past and present. London, Routledge.

KALICZ, N. 1998
Figürliche Kunst und bemalte Keramik aus dem Neolithikum Westungarns. Budapest, Archaeolingua, Series Minor 10.

KALICZ, N. – MAKKAY, J. 1977
Die Linienbandkeramik in der Grossen Ungarischen Tiefebene. Budapest, Akadémiai Kiadó.

KALICZ, N. – RACZKY, P. 1984
Preliminary report on the 1977-1982 excavation at the Neolithic and Bronze Age settlement of Berettyóújfalu-Herpály. Part I: Neolithic. *Acta Archaeologica Hungarica* 36, 85-136.

KALICZ, N. – RACZKY, P. 1987
The Late Neolithic of the Tisza region. A survey of recent archaeological research. In: Raczky, P. (ed.), *The Late Neolithic of the Tisza region.* Budapest-Szolnok, Szolnok County Museums, 11-30.

KOREK, J. 1987
Szegvár-Tűzköves. In: Raczky, P. (ed), *The Late Neolithic in the Tisza region.* Budapest-Szolnok, Szolnok County Museums, 47-60.

KOREK, J. 1989
Die Theiss-Kultur in der mittleren und nördlichen Theissgegend. Inventaria Praehistorica Hungariae 3, Budapest, Magyar Nemzeti Múzeum.

KOSSE, K. 1979
Settlement ecology of the Körös and Linear Pottery cultures in Hungary. International Series 64. Oxford: BAR.

KOSSINNA, G. 1896
Die vorgeschichtliche Ausbreitung der Germanen in Deutschland. *Zeitschrift des Vereins für Volkskunde* 6, 1-14.

KOSSINNA, G. 1911
Die Herkunft der Germanen. Zur Methode der Siedlungsarchäologie. Mannus-Bibliothek 6. Würzburg, Kabitzsch.

KÜCHLER, S. 1994
Landscape as memory. In: Bender, B. (ed.), *Landscape: politics and perspectives*. Oxford, Berg, 85-106.

LASZLOVSZKY, J. – SIKLÓDI, CS. 1990
Theoretical archaeology without theory. In: Hodder, I. (ed.), *Archaeological theory in Europe: the last three decades*. London, Routledge, 272-298.

LEDERMANN, R. 1990
Contested order: gender and society in the Southern New Guinea Highlands. In: Sanday, P. R., Goodenough, R. G. (eds.), *Beyond the second sex: new directions in the analysis of gender*. Philadelphia, University of Philadelphia Press, 45-73.

LESICK, K. S. 1996

Re-engendering gender: some theoretical and methodological concerns on a burgeoning archaeological pursuit. In: Moore J., Scott, E. (eds.), *Invisible people and processes*. Writing gender and children into European archaeology. Leicester, Leicester University Press, 31-41.

LICHARDUS, J. 1988

Der westpontische Raum und die Anfänge der kupferzeitlichen Zivilisation. In: Fol, A., Lichardus, J. (eds.), *Macht, Herrschaft und Gold*. Saarbrücken, Moderne Galerie des Saarland Museums, 79-130.

LICHARDUS, J. 1991

Die Kupferzeit als historische Epoche. Versuch einer Deutung. In: Lichardus, J. (ed.), *Die Kupferzeit als historische Epoche*. Saarbrücker Beiträge zur Altertumskunde 55. Saarbrücken, Saarland Museum, 763-800.

LUCY, S. J. 1998

The Early Anglo-Saxon cemeteries of East Yorkshire. An analysis and reinterpretation. British Series 272. Oxford, BAR.

Magyar Nemzeti Múzeum 1991

Lithoteka

Magyar Nemzeti Múzeum 1999

M-3 autópálya. CD ROM. Budapest, Nemzeti Múzeum.

MAKKAY, J. 1969

The Late Neolithic Tordos group of signs. *Alba Regia* 10, 9-49.

MAKKAY, J. 1975

Über neolithische Opferformen. In: Anati, E. (ed.), *Les religions de la préhistoire*. Capo di Ponte, Brescia, Centro Camuno di Studi Preistorici, 161-173.

MAKKAY, J. 1982

A magyarországi neolitikum kutatásának új eredményei. Budapest, Akadémiai Kiadó.

MAKKAY, J. 1982a

Some comments on the settlement patterns of the Alföld Linear Pottery. In: Pavuk, J. (ed.), *Siedlungen der Kultur mit Linearkeramik in Europa*. Nitra, Slowakische Akademie der Wissenschaften, 157-166.

MAKKAY, J. 1983

Foundation sacrifices in Neolithic houses of the Carpathian Basin. In: Anati, E. (ed.), *The intellectual expressions of prehistoric art and religion*. Capo di Ponte, Brescia: Centro Camuno di Studi Preistorici, 157-167.

MAKKAY, J. 1983a
Metal forks as symbols of power and religion. *Acta Archaeologica Hungarica* 35, 313-344.

MAKKAY, J. 1985
Diffusionism, antidiffusionism and chronology: some general remarks. *Acta Archaeologica Hungarica* 37, 1-12.

MAKKAY, J. 1992
Excavations at the Körös culture settlement of Endrőd-Öregszőlők 119. In: Bökönyi, S. (ed.), *Cultural and landscape changes in South East Hungary I. Reports on the Gyomaendrőd Project*. Budapest, Archaeolingua, 121-194.

MANN, M. 1986
The sources of social power. Volume 1. A history of power from the beginning to A. D. 1760. Cambridge: Cambridge University Press.

MANN, M. 1993
The sources of social power. Volume 2. A history of power from A. D. 1760 to AD 1914. Cambridge: Cambridge University Press.

MEIER-ARENDT, W. – RACZKY, P. (eds.) 1990
Alltag und Religion. Jungsteinzeit in Ost-Ungarn. Frankfurt am Main, Museum für Vor- und Frühgeschichte.

MEISENHEIMER, M. 1989
Das Totenritual, geprägt durch Jenseitsvorstellungen und Gesellschafts-realität: Theorie des Totenrituals eines Kupferzeitlichen Friedhofs zu Tiszapolgár-Basatanya (Ungarn). International Series 475. Oxford: BAR.

MILLER, D. 1985
Artefacts as categories. A study of ceramic variability in Central India. Cambridge, Cambridge University Press.

MILLER, D. 1987
Material culture and mass consumption. Oxford, Blackwell.

MILLER, D. – TILLEY,C. (eds.) 1985
Ideology, power and prehistory. Cambridge, Cambridge University Press

MILLER, D. – ROWLANDS, M. – TILLEY, C. (eds.) 1991
Domination and resistance. London, Routledge.

MIZOGUCHI, K. 1993
Time in the reproduction of mortuary practices. *World Archaeology* 25/2, 223-235.

MOORE, H. 1993
The differences within and the differences between. In: del Valle, T. (ed.), *Gendered Anthropology*. London, Routledge, 193-204.

MUNN, N. 1973
Walbiri iconography: graphic representation and cultural symbolism in a central Australian society. Ithaca, New York, Cornell University Press.

MUNN, N. 1986
The fame of Gawa. A symbolic study of value transformation in a Massim (Papua New Guinea) society. London, Duke University Press.

NANDRIS, J. 1972
Relations between the Mesolithic, the First Temperate Neolithic and the Bandkeramik: the nature of the problem. *Alba Regia* 6, 61-70.

NEMESKÉRI, J.– SZATHMÁRI, L. 1987
An anthropological evaluation of the Indo-European problem: the anthropological and demographic transition in the Danube Basin. In: Skomal, S. N., Polomé, E. C. (eds.), *Proto-Indo-European. The archaeology of a linguistic problem.* Studies in honor of Marija Gimbutas. Washington, D. C., Institute for the Study of Man, 88-121.

O'SHEA, J. 1984
Mortuary variability: an archaeological investigation. London, Academic Press.

O'SHEA, J. 1996
Villagers of the Maros: a portrait of an early Bronze Age society. New York, Plenum Press.

PARKER PEARSON, M. 1999
The archaeology of death and burial. Stroud, Alan Sutton.

PATAY, P. 1974
Die hochkupferzeitliche Bodrogkeresztúr-Kultur. *Bericht der Römisch-Germanischen Kommission* 55, 1-71.

RACZKY, P. (ed.) 1987
The Late Neolithic in the Tisza region. Budapest-Szolnok, Szolnok County Museum.

RACZKY, P. 1987a
Öcsöd-Köveshalom. A settlement of the Tisza culture. In: Raczky, P. (ed.), *The Late Neolithic in the Tisza region.* Budapest-Szolnok, Szolnok County Museum, 61-83.

RACZKY, P. 1988
A Tisza-Vidék Kulturális és kronológiai kapcsolatai a Balkánnal és az égeikummal a neolitikum, rézkor időszakában. Szolnok, Szolnok County Museum.

RACZKY, P. 1994
Two late neolithic 'hoards' from Csóka (Coka)-Kremenyak in the Vojvodina. In: G. Lőrinczky (ed.), *A kőkortól a középkorig. Tanulmányok*

Trogmayer Ottó 60. Születésnapjára. Szeged, Móra Ferenc Múzeum, 161-172.

RACZKY, P. 1997
Polgár-Csőszhalom-dűlő. In: Raczky, P., Kovács, T., Anders, A. (eds.), *Utak a múltba. Az M3-as autópálya Régészeti leletmentései.* Budapest Hungarian National Museum and Archaeological Institute, ELTE, 34-43.

RACZKY, P. 1998
The late neolithic tell of Polgár-Csőszhalom and its relationship to the external horizontal settlement in light of recent archaeological data. In: Anreiter, P., Bartosiewicz, L., Jerem, E., Meid, W. (eds.), *Man and the Animal World.* Budapest, Archaeolingua, 481-489.

RACZKY, P. – CZAJLIK, Z. – MARTON, A. – HOLL, B. – PUSZTA, S. 1997
GIS and the evaluation of rescue excavations along the M3 motorway in Hungary. *Porocilo* 224, 157-170.

RACZKY, P. – MEIER-ARENDT, W. – HAJDÚ, ZS. – KURUCZ, K. – NAGY, E. 1996
Two unique assemblages from the Late Neolithic tell settlement at Polgár-Csőszhalom. In: Kovács, T. (ed.), *Studien zur Metallindustrie im Karpatenbecken und den benachbarten Regionen.* Budapest, Magyar Nemzeti Múzeum, 17-30.

RACZKY, P. – MEIER-ARENDT, W. – KURUCZ, K. – HAJDÚ, ZS. – SZIKORA, A. 1994
A Late Neolithic settlement in the Upper Tisza region and its cultural connections (Preliminary report). *A Jósa András Múzeum Évkönyve* 36, 231-240.

RACZKY, P. – SELEANU, M. – RÓZSA, G. et al.1985
Öcsöd-Köveshalom. The intensive topographical and archaeological investigation of a Late Neolithic site. Preliminary report. *Mitteilungen des Archäologisches Instituts* 14, 251-278, Budapest.

REGA, E. 1997
Age, gender and biological reality in the Early Bronze Age cemetery at Mokrin. In: Moore, J., Scott, E. (eds.), *Invisible people and processes. Writing women and children into European archaeology.* Leicester, Leicester University Press, 229-247.

REGENYE, J. 1994
Építési áldozat a lengyeli kultúrából, Bakonyszűcsről. In: Lőrinczy, G. (ed.), *A kőkortól a középkorig. Tanulmányok Trogmayer Ottó 60. Születésnapjára.* Szeged, Móra Ferenc Múzeum, 151-160.

RENFREW, C. 1978
Varna and the social context of early metallurgy. *Antiquity* 52, 199-203.

RENFREW, C. 1986

Varna and the emergence of wealth. In: Appadurai, A. (ed.), *The social life of things*. Cambridge, Cambridge University Press, 141-68.

ROBB, J. E. 1994

Burial and social reproduction in the Peninsular Italian Neolithic. *Journal of Mediterranean Archaeology* 7/1, 27-71.

RORTY, R. 1989

Contingency, irony and solidarity. Cambridge, Cambridge University Press.

SAHLINS, M. 1972

Stone Age economics. Chicago, Aldine.

SAXE, A. A. 1970

Social dimensions of mortuary practices. Ph. D. thesis. University of Michigan, Ann Arbor.

SCHÜLKE, A. 1999

On Christianisation and grave finds. *European Journal of Archaeology* 2/1, 77-106.

SHANKS, M. – TILLEY, C. 1987

Social theory and archaeology. Cambridge, Polity Press.

SHERRATT, A. 1979

Plough and pastoralism: aspects of the secondary products revolution. In: Hodder, I., Hammond, N., Isaac, G. (eds.), *Patterns in the past*. Cambridge, Cambridge University Press, 261-305.

SHERRATT, A. 1982

Mobile resources: settlement and exchange in early agricultural Europe. In: Renfrew, C., Shennan, S. J. (eds.), *Ranking, resources and exchange*. Cambridge, Cambridge University Press, 13-26.

SHERRATT, A. 1982a-1983

The development of Neolithic and Copper Age settlement in the Great Hungarian Plain: Part 1. The regional setting, and Part 2. Site surveys and settlement dynamics. *Oxford Journal of Archaeology* 1/3, 287-316; 2/1, 13-41.

SHERRATT, A. 1987

Neolithic exchange systems in Central Europe, in Sieveking. In: G. de G., Newcomer, M. N. (eds.), *The human uses of flint and chert*. Brighton, Harvester Press, 193-204.

SKOMAL, S. 1980

The social organisation of the Tiszapolgár Group at Basatanya-Carpathian Basin Copper Age. *Journal of Indo-European Studies* 8/1, 75-91.

SOFAER DEREVENSKI, J. 1997
Age and gender at the site of Tiszapolgár-Basatanya, Hungary. *Antiquity* 71, 875-889.

SOFAER DEREVENSKI, J. 2000
Rings of life: the role of early metalwork in mediating the gendered life-course. *World Archaeology* 31/3, 389-406.

SOPRONI, S. 1956
Budakalász, Luppa-csárda. In: Banner, J. (ed.), *Die Peceler Kultur.* Archaeologia Hungarica 35, Budapest, Akadémiai Kiadó, 111-128.

SØRENSEN, M. L. S. 1997
Reading dress: the construction of social categories and identities in Bronze Age Europe. *Journal of European Archaeology* 5/1, 93-114.

SPECTOR, J. 1991
What this awl means: towards a feminist archaeology. In: Gero, J.M., Conkey, M.W. (eds), *Engendering archaeology. Women and prehistory.* Oxford, Blackwell, 388-405.

SPINDLER, K. 1994
The man in the ice. London. Weidenfeld & Nicolson.

SPRIGGS, M. (ed.) 1984
Marxist approaches in archaeology. Cambridge, Cambridge University Press.

TAINTER, J. 1978
Mortuary practices and the study of prehistoric social systems. *Advances in Archaeological Method and Theory* 1, 105-141.

TAKÁCS, I. 1992
The fish bones. In: Bökönyi, S. (ed.), *Cultural and landscape changes in South East Hungary I. Reports on the Gyomaendrőd Project.* Budapest, Archaeolingua, 301-311.

TORMA, I. 1969
A veszprém megyei résgészeti topográfiai kutatások őskori vonatkozású eredményeiről. (Über vorgeschichtliche Ergebnisse der archäologischen Topographie auf dem Gebiet des Komitats Veszprém) *Veszprém Megyei Múzeumok Közleményei* 8, 75-81.

TRINGHAM, R. 1991
Households with faces: the challenge of gender in prehistoric architectural remains. In: Gero, J. M., Conkey, M. W. (eds.), *Engendering archaeology.* Oxford, Blackwell, 93-131.

WHITTLE, A. 1996
Europe in the Neolithic. The creation of new worlds. Cambridge, Cambridge University Press.

WHITTLE, A. 1998
Beziehungen zwischen Individuum und Gruppe: Fragen zur Identität im Neolithikum der ungarischen Tiefebene. *Ethnographisch-Archäologische Zeitschrift* 39, 465-487.

WHITTLE, A. 1998a
Fish, faces and fingers: presences and symbolic identities in the Mesolithic-Neolithic transition in the Carpathian Basin. *Documenta Praehistorica* 25,133-150.

WYATT, S. 1994
The dead of Tiszapolgár-Basatanya: a Copper Age cemetery in the Great Hungarian Plan. Unpublished M. A. (Hons) dissertation, University of Edinburgh.

ZALAI-GAÁL, I. 1986
Mórágy-Tűzkődomb: Entwurf sozialarchäologische Forschungen. In: Chropovsky, B., Friesinger, H. (eds.), *Internationales Symposium über die Lengyel-Kultur.* Nove Vozokaný. Nitra-Wien, Slovakische Akademie der Wissenschaften, 333-338.

ZALAI-GAÁL, I. 1986a
Sozialarchäologische Forschungsmöglichkeiten aufgrund der spät-neolithischen Gräbergruppen in Südwestlichem Ungarn. *A Béri Balogh Ádám Múzeum Évkönyve* 13, 139-154.

ZALAI-GAÁL, I. 1988
Sozialarchäologische Untersuchungen des mitteleuropäischen Neolithikums aufgrund der Gräberfeldanalyse. *A Béri Balogh Ádám Múzeum Évkönyve* 14.

ZALAI-GAÁL, I. 1996
Die Kupferfunde der Lengyel-Kultur im südlichen Transdanubien. *Acta Archaeologica Hungarica* 48, 1-34.